EVALUATION and OPTIMIZATION

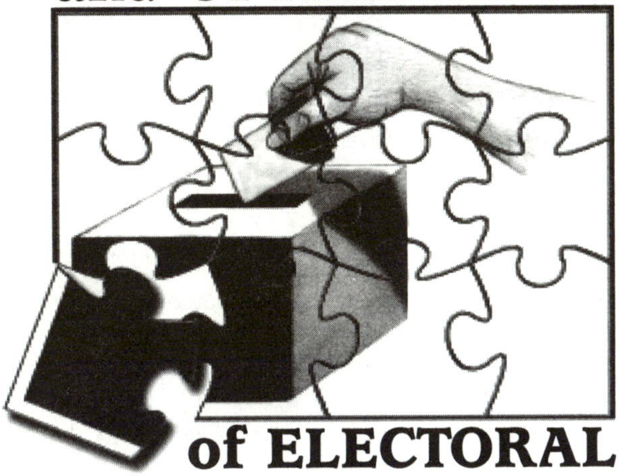

# of ELECTORAL SYSTEMS

SIAM Monographs on Discrete Mathematics and Applications

The series includes advanced monographs reporting on the most recent theoretical, computational, or applied developments in the field, introductory volumes aimed at mathematicians and other mathematically motivated readers interested in understanding certain areas of pure or applied combinatorics, and graduate textbooks. The volumes are devoted to various areas of discrete mathematics and its applications.

Mathematicians, computer scientists, operations researchers, computationally oriented natural and social scientists, engineers, medical researchers and other practitioners will find the volumes of interest.

Series Volumes

Grilli di Cortona, P., Manzi, C., Pennisi, A., Ricca, F., and Simeone, B., *Evaluation and Optimization of Electoral Systems*

EVALUATION and OPTIMIZATION

# of ELECTORAL SYSTEMS

Pietro Grilli di Cortona
University of Rome 3
Rome, Italy

Cecilia Manzi
National Insurance Institute
Rome, Italy

Aline Pennisi
University of Rome "La Sapienza"
Rome, Italy

Federica Ricca
University of Rome "La Sapienza"
Rome, Italy

Bruno Simeone
University of Rome "La Sapienza"
Rome, Italy

siam.
Society for Industrial and Applied Mathematics Philadelphia

10 9 8 7 6 5 4 3 2 1

Library of Congress Cataloging-in-Publication Data

Evaluation and optimization of electoral systems / Pietro Grilli di Cortona ... [et al.].
p. cm. -- (SIAM monographs on discrete mathematics and applications)
Includes bibliographical references and index.
ISBN 0-89871-422-2 (pbk.)
1. Elections--Mathematical models. I. Grilli Di Cortona, Pietro. II. Society for Industrial and Applied Mathematics. III. Series.
JF1001.E93 1998
324.6'01'51--dc21 98-39941

To our loved ones:

Barbara, Bernardo, Giovanni, Sofia;

Raffaele, Carla;

Maman, Papà, Papà Pierre, Mamie Claude;

Mamma, Papà, Paola e Marco, Andrea;

Giusi, Giulio, Chiara, Maria.

Contents

Preface

> Scientific and humanist approaches are not competitive but supportive, and both are absolutely necessary.
>
> Robert C. Wood

The choice of an appropriate electoral system, capable of guaranteeing a representative, stable, and efficient government, has been a major concern of the political debate in many countries in the last few years.

But how can different electoral systems be defined? What are the criteria used to determine the "quality" of an electoral system, and how can we measure them? And, again, how can we use the different criteria to design an electoral system? Moreover, how can we measure the trade-offs between conflicting criteria?

This book is the result of the contribution of experts from different fields, such as political sciences, operations research, statistics, and decision sciences. It tries to answer the questions above offering a general methodology and a set of mathematical tools, many of which are original, in order to analyze and design electoral systems.

The analysis of electoral systems is generally thought of as a matter of political and social disciplines. Nevertheless, it is possible to analyze electoral systems from a mathematical point of view. Quantitative approaches to the study of electoral systems already exist and have evolved in different directions:

- an *axiomatic* approach, developed in Nurmi (1987) and Balinski and Young (1982a);
- a *statistical* approach, found in Key (1954) and in the recent Italian volume by Vitali (1995);
- a *game theory* approach, for example, in Brams (1975) and Ordeshook (1992);
- a *geometric approach* developed in Saari (1994).

In this book we propose still another quantitative point of view: the optimization of one or many criteria.

The main steps we follow are

(1) the decomposition of the electoral process into phases;

(2) the definition of a general formal model for electoral systems;

(3) the identification of qualitative criteria assessing the performance of an electoral system;

(4) the definition of appropriate performance indicators;

(5) the optimization of a single criterion and comparison between optimal solutions for different single criteria;

(6) multicriteria optimization and trade-offs between criteria.

Part I develops steps 1–4, Part II investigates step 5, and Part III analyzes step 6.

In Chapter 1 the electoral process is viewed as a whole. Particular attention is given to the electoral system. The heart of the electoral system is the electoral formula which transforms party votes into legislature seats. It is also possible to consider upstream and downstream elements in this process, for example, those affecting the voters' decisions (constitutional rules, political situation) and those affected by the seat apportionment (party coalitions, creation of a government).

In fact, every electoral process for the formation of a government can be decomposed into four phases (each one affecting the next) from the votes cast by the voters. We also take into consideration feedback effects.

A general model of the electoral system is developed using the elementary concepts of set theory (Chapter 2). This model provides clear definitions and a unified approach to the theory of electoral systems. We did not find a comprehensive model of this sort in the literature.

The usefulness of a mathematical model largely depends on the precision with which it represents the real world. The electoral systems currently used in many different countries are all obtainable as particular cases of the model we propose. A by-product of this model is, in fact, an original coding scheme which can be used to identify and classify electoral systems, similar to the one commonly used in scheduling problems.

Chapter 3 deals with the problem of identifying appropriate criteria to define the *quality* of an electoral system. Our formulation of the problem is certainly influenced by the idea of "total quality" (and in particular the book written by Vacca, 1994). In any case, the aim is to provide an organized review of the existing criteria and to propose or develop indexes to compare electoral systems on the basis of such criteria. Finally, we can divide the indexes with respect to the phases identified above into three categories:

- indexes that describe the political system before the vote-seat transformation (such as electoral participation, party fragmentation, and electoral fluctuation);

- indexes that explain what happens during the vote-seat transformation (in particular, proportionality);
- indexes that evaluate what happens after the vote-seat transformation (such as government stability and coalition power of the parties).

Part II is devoted to the core of every electoral system, the electoral formulas. In Chapter 4 we review the axiomatic approach. A minimal list of reasonable properties that every electoral formula should satisfy is considered. Unfortunately, just as stated in the electoral counterpart of Arrow's impossibility theorem, there exists no perfect electoral system, that is, no system that simultaneously satisfies such a list of properties.

In Chapter 5 we recall the most important properties satisfied by the most common proportional and majoritarian systems. In Chapter 6 we give a procedural description of an electoral formula. In fact most electoral formulas are not easily described by closed-form analytical expressions, but they operate more like *algorithms* that receive an input (the vote distribution) and generate an output (a particular seat assignment) through a successive sequence of elementary operations. In our view, each electoral formula is a specific algorithm designed to minimize a particular cost function. In particular, proportional formulas are efficient algorithms used to solve an integer optimization problem with specific objective functions, some of which are illustrated in detail. The objective functions, which are different according to each formula, represent different proportionality indexes. They can be viewed as the "hidden" criteria on which each formula is based. We think that this kind of approach can give a theoretical background to the well-known proportional formulas. In fact, this will allow us to point out the weakness of common methodologies in the analysis of proportional systems and to introduce the concept of Schur-convex functions as a relevant class of disproportionality indexes. In fact, Schur-convexity had a key role in the beginning of this century, when the pioneers of welfare economics—and in particular Lorenz, Pigou, and Dalton—were concerned with comparing different income or wealth distributions. We use them in the electoral context to measure the inequality of the seat-to-vote ratios in the legislature apportionment.

We also show that, for a significant class of disproportionality indexes, inequality of the seat-to-vote ratios can be minimized through very simple *greedy* or *exchange* algorithms: the former ones allocate one seat at a time so as to obtain at each step the largest possible decrease of the chosen index; the latter ones, starting from an arbitrary seat allocation, perform a transfer of one seat from a party to another every time such transfer makes the index decrease.

In Chapter 8 we discuss mixed electoral systems. We are particularly interested in analyzing the effects of the variation of the quota of seats assigned by a majoritarian formula (with respect to those assigned with a proportional system) on electoral campaign strategies and electoral outcomes.

The methodology presented in this part of the book cannot tell the reader how to design or choose the "best" electoral system. However, it offers a rich background which is necessary to explain the transformation of votes into seats and to understand why, although all parties are equal, "some are more equal than others."

In Part III we are interested in designing electoral districts. The choice of an appropriate district plan is no less important than that of an appropriate electoral formula, even though districting has not been an issue in the political debate of many countries. Choosing criteria necessary to plan districts is certainly a technical problem, but districting cannot be considered merely a technical issue because it has real political consequences. In Chapter 9 we present some of the traps that may be hidden behind the planning of electoral districts and how district plans may be used to manipulate electoral outcomes favoring some political party (gerrymandering).

In particular, we will show that, while the size of the districts together with the electoral formula certainly have an impact on the seat distribution, the shape of the electoral districts can drastically affect the outcome.

District planning must be based on "neutral" criteria, that is, criteria that are independent of the past or the expected vote distribution, so as to prevent the possibility of manipulation. In Chapter 10 we review the most important of these criteria and in Chapter 11 we present the correspondent indexes. Chapter 12 deals with traditional and more recent combinatorial optimization algorithms used to find district plans based on the above criteria. Finally, we discuss trade-offs between districting criteria, illustrating them in the concrete case of the Italian district plan used for Parliament seats in the elections of March 27, 1994 and April 21, 1996. The three main criteria considered in this analysis are population equality, compactness, and conformity to administrative boundaries. What happens when we optimize with respect to only one of the possible criteria? Generally we would assume that the values obtained for the other criteria worsen, but this is not always the case. In fact a careful analysis of the trade-offs between criteria is full of surprises. For example, is it possible to design the Italian electoral districts in such a way that all three criteria present simultaneously better values? The answer to this question is in section 12.6.

Part IV presents some observations by the political scientist of our team on the benefits and on the limits of the methodology used to analyze and design electoral systems in this book. This part is written from a completely different point of view, focusing on the *political* aspects of electoral reforms. Some recent examples will be analyzed.

To summarize, some original contributions the reader will find in this book are the following:

- a general formal model to describe electoral systems;
- a coding system to identify and classify electoral systems;
- a review, classification, and analysis of many traditional and new performance indicators;
- electoral formulas are viewed as algorithms to minimize actual cost functions or proportionality indexes;

- a quantification of the effects of mixed systems on the electoral outcome;
- the presentation of automatic procedures for electoral districting based on neutral criteria;
- the explanation of the electoral reform process in terms of the transformations occurring in the political party structure.

This book will not give the reader a recipe for the best electoral system. It does not intend to give final answers, but it invites the reader to look at electoral systems from different viewpoints and to use results that can be obtained through mathematical models.

Mathematical models are powerful instruments. They can be used like a "microscope" to focus on the different interconnecting components of a complex system and like a "telescope" to explore (with the aid of a computer) a huge number of alternative solutions. The methodological tools presented in this book are useful for the analysis of difficult decision problems. They should be used together with other methodologies to obtain a better understanding of real situations. In fact, the computational efficiency provided by a mathematical model together with the judgmental capabilities, the experience, and the intuition of a human decision maker are the key to success in the solution of many difficult decision problems.

Although this book has been thoroughly discussed by its five authors, Cecilia Manzi was responsible for Part I, Aline Pennisi for Part II, and Federica Ricca for Part III. Pietro Grilli di Cortona is the author of Part IV, "The Process of Electoral Reform: A Retrospective Critical View of a Political Scientist." Bruno Simeone is the main architect of this volume and the coordinator of the workgroup. Many of his ideas emerge through the first three parts of this book. The presentation of many paragraphs within these parts was strongly influenced by him.

The idea of writing such a book took place during the ADEN Project, aimed to developing interactive software tools for district planning, which was carried out by Professor Mario Lucertini (University of Rome "Tor Vergata"), Dr. Fabio Arcese (Ufficio Studi della Camera), and Drs. Maria Gloria Battista and Orazio Biasi (Italsiel), together with Bruno Simeone. We thank them all for the many discussions which have strongly affected Part III. We also thank Dr. Mario Salani for his useful suggestions during the early stages of the preparation of the manuscript. Finally, special thanks to Professor Peter Hammer, who encouraged us to submit our book for publication in the SIAM Monographs on Discrete Mathematics and Applications, and to SIAM staff.

We have made every possible effort to offer an accessible exposition to readers who are interested in electoral systems but are not professional mathematicians. Those who are not concerned about mathematical details may skip through theorems, or at least their proofs, since we have always tried to provide a nontechnical explanation of their meaning and consequences. The writing style we have adopted is descriptive and informal, and often we have purposefully sacrificed terseness in favor of readability, at the cost of being, sometimes, repetitive or redundant. Mathematicians will forgive us for this. In fact, this volume is intended for both political scientists and mathematicians. The former will find an unusual but stimulating approach presented in a simple and clear manner and the latter will see familiar tools used in a real-life context of great topical interest.

Pietro Grilli di Cortona
Cecilia Manzi
Aline Pennisi
Federica Ricca
Bruno Simeone

Part I

Classification and Evaluation of Electoral Systems

The beginning of wisdom is the definition of terms.
Socrates

The design of electoral systems is usually left to political and social scientists, although many mathematical aspects are involved: a formal interpretation of the electoral process can be useful to evaluate its performance. In our opinion, the debate on electoral systems can be significantly enlightened by a quantitative approach which captures their technical properties and the interrelationships between their different components.

The first part of this volume attempts a systematic analysis of the components of the electoral process by subdividing it into four distinct phases. In fact, in spite of the numerous works which have flourished in the literature, the many theories consolidated throughout time, and the historical concern for this topic, the analysis of electoral systems has never been developed on a homogeneous formal background. For example, the major interest for plurality and majority-like systems in the development of the theory of social choice has led them to be examined in a totally different way with respect to proportional systems, to the point that the comparison between these two approaches is merely confined to the examination of their political and historical development. Interdisciplinary approaches are very rare.

Nevertheless, a formal approach to the electoral process must be based on the use of the same notation, tools, and language. The classification of electoral systems must be based on common criteria, and these criteria should be used to measure the *quality* of electoral systems, as well as to provide an objective basis for their evaluation. This is the main aim of the following paragraphs.

Chapter 1

The Four Phases of an Electoral Process

Electoral systems are designed to make the representation of the different and often conflicting interests of a multitude of citizens possible through the establishment of a public assembly, usually called Parliament, and a government. To describe an electoral system we must refer to its many components, such as the definition of the electoral body, the size of the legislation, the rules regulating the electoral campaign, the districting procedure, and of course the formula adopted to translate votes into seats. However, this last component, which is often identified as the electoral system itself, is not enough to explain electoral outcomes.

Every component of the electoral system has a fundamental role and can be used to modify its results. A simple but effective representation of the electoral system must consider the whole process, starting from the preferences expressed by the voters to the establishment of a government. The following four phases can be pointed out:

(1) definition of the electoral rules;

(2) the vote;

(3) the translation of votes into seats;

(4) the government formation.

The main ingredients in every electoral system are the parties, the candidates, the votes, and the number of available seats. These can be viewed as the input data of a model whose solution produces an output in terms of the seat distribution assigned to the different parties.

Now, what are the factors that determine the input, and what are the consequences of the output?

A diagram can be used to point out how many different elements interact in the electoral process (Figure 1.1).

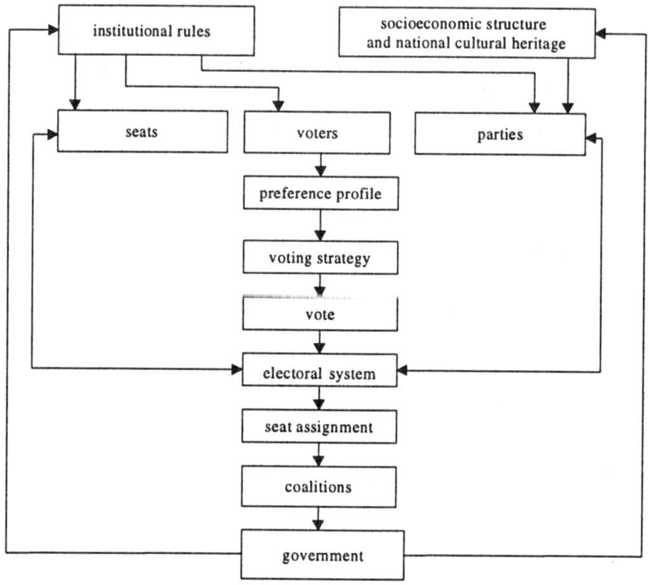

Fig. 1.1. *The electoral process.*

Definition of Electoral Rules

The institutional rules and the social and political structure of a country are the main elements that determine the number of parties and the competition between them. Institutional rules also define the number, size, and shape of the electoral districts, that is, which and how many territorial aggregates are to be considered, how many voters are assigned to each aggregate, and how many seats will be elected in each aggregate. This phase must therefore be considered as a preliminary and preparatory phase (Figure 1.2).

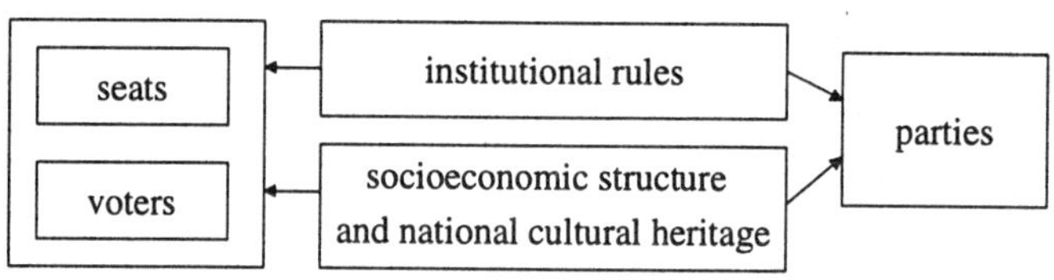

Fig. 1.2. *Phase 1: the definition of electoral rules.*

The Vote

Once the candidates and the different parties are established, each voter can build his or her own preference profile, i.e., a ranking in which all the alternatives are ordered from the most preferable to the less preferred one. It must be noticed that the voters might not necessarily vote for their most preferred alternative, or, in other words, their vote is not necessarily *sincere*. In the meantime, in fact, each voter will study a voting strategy, based on the expected behavior of the other voters, and a final decision (that is, the ballot cast) will be taken (Figure 1.3).

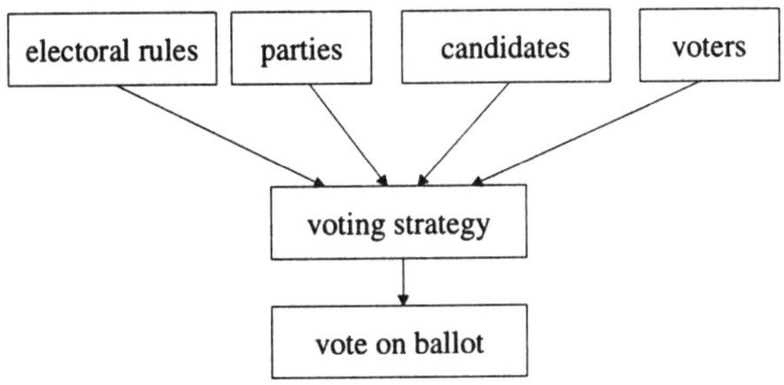

Fig. 1.3. *Phase 2: the vote.*

The Translation of Votes into Seats

Once the inputs have been defined, that is, the parties, the electoral body, and the seats at stake, and once the votes have been cast, the crucial phase of translating votes into seats starts. The rule adopted is described by the *electoral engine*—including one or more electoral formulas—which essentially constitutes the third phase of the process (Figure 1.4).

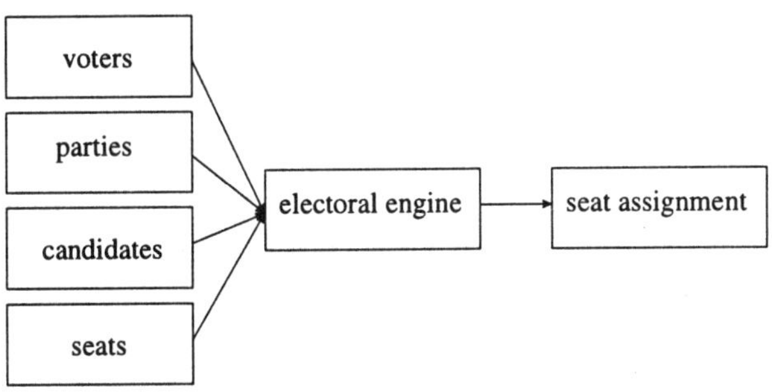

Fig. 1.4. *Phase 3: the translation of votes into seats.*

Government Formation

However, once the seats have been assigned, the process is not completed. In fact, the electoral outcome can produce many different alternatives before the establishment of a government: in most cases it is necessary that several parties join the same coalition so that at least 51% of the seats are represented.

This last phase of the electoral process has deep consequences on the whole system. There is a *feedback* effect on the other phases of the system, and on the first phase in particular, since the Parliament is the institution that will define the rules of the game to be played in the future. The consequences of the electoral system directly act on all the components mentioned up to now, from the behavior of the voters to the party system (Figure 1.5).

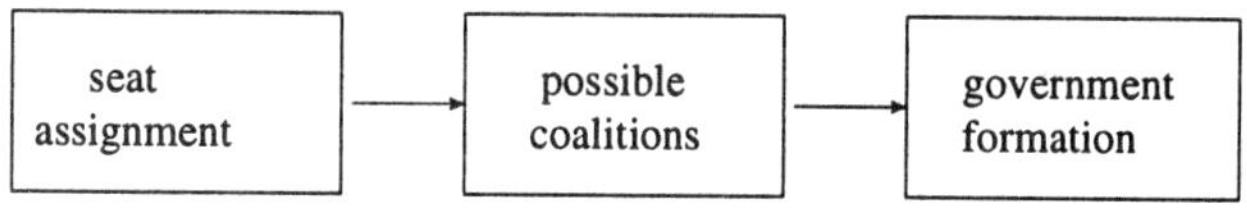

Fig. 1.5. *Phase 4: government formation.*

The four phases we have described are therefore inextricably bound and the scheme we have proposed is dynamic, since they act in a continuous cycle. The final result (the establishment of a government) can therefore be thought of as a result of all of the previous components of the process. The government depends on the possible coalitions, which depend on the seat distribution among the parties, which in turn depend on the electoral formula and the votes cast according to the preference profiles and strategies of the voters. Moreover, the party system is determined by the institutional rules, by the consequences of the electoral formula, and by the historical, social, and cultural background of the country.

Chapter 2

A Unified Description of Electoral Systems

In this chapter we intend to formally describe what an electoral system is. The general opinion is that the notion of an electoral system is so widespread it need not be explained, to the point that the media often discuss electoral reform without mentioning the basic elements on which a reform can act. An electoral system is usually a quite complicated device to describe, and a systematic description should start by listing the elementary factors that are involved and that deserve a discussion on their own. If we were to ask for a brief definition of an electoral system, a simple answer could be that an electoral system is the set of rules that allow us to select candidates. This is substantially correct, but it automatically leads us to ask which rules allow us to select these candidates. Specifying the rules means talking about type and number of ballots, size of the legislature, computational rules, size of the districts, number of parties, and candidates, etc., as we have already tried to point out in the previous chapter, where the electoral process was split into four phases.

Different combinations of these rules give rise to a surprisingly large number of different electoral systems: hundreds of them are currently in use, and undoubtedly many others are conceivable and no less sound than the existing ones.

Due to the intrinsic complexity of electoral systems, building a general and comprehensive paradigm is no easy task. Were one to write down a computer program for generating all possible electoral systems, one would definitely need a general formal model describing their components and the connections between different components. Some researchers have gone a long way in this direction (Nurmi's (1987) voting systems and Cox's (1997) formulaic matrices are particularly worthy of mention), but, as far as we know, a general formal model is still lacking in the literature.

To describe such a model, we advocate the use of the language of elementary set theory for the following reasons:

(1) it is a *powerful* and *universal* language for describing entities and relations;

(2) once primitive sets (voters, candidates, parties, seats) have been defined, the other components of the electoral system (districts, lists, ballots, seat assignments, and so on) can be derived in a *simple* and *natural* way through the use of concepts of set theory (subsets, mappings, partitions, coverings, etc.);

(3) it leads to *unambiguous* and crystal-clear definitions of the objects under consideration.

In the following section we will therefore identify and examine the building blocks of every electoral system (its "primitive elements") and the relationships that link them to one another.

2.1 A General Formal Model

The building blocks of an electoral system can be grouped into four distinct classes of elementary units: the voters, the candidates, the parties, and the seats. They can be related among themselves in different ways. In fact, candidates can be aggregated from the point of view of the parties they belong to or according to the districts in which they are competing. Electors can be aggregated in different ways too: according to the party they vote for or from the point of view of the districts they belong to. Parties can be aggregated into coalitions, districts can be aggregated into regions, and so on. In formal terms, we can identify partitions and coverings on the set of candidates and on the set of electors.

Each elementary unit is, therefore, part of a set as follows:

- $\mathbf{V}$, the set of voters (or electors) with cardinality $|\mathbf{V}| = V$;
- $\mathbf{C}$, the set of candidates with cardinality $|\mathbf{C}| = C$;
- $\mathbf{I}$, the set of parties with cardinality $|\mathbf{I}| = n$;
- $\mathbf{S}$, the set of seats with cardinality $|\mathbf{S}| = S$.

In order to carry out the computational rules that form an electoral system, the set of voters $\mathbf{V}$ is usually not considered altogether. Mainly for organizational purposes, but also for the philosophical notion of representation underlying the different systems, the voters are divided into a certain number k of "small" or manageable groups of citizens belonging to the same geographic area, called political *districts*. This means that each political district corresponds to a subset of voters, and each voter belongs only to one political district at a time. The subset of voters corresponding to the jth district can be denoted by V_j. The political districts define a partition $\pi = (V_1, V_2, \ldots, V_k)$ of the set of voters, so that

$$\bigcup_{j=1}^{k} V_j = \mathbf{V}$$

and

$$V_i \bigcap V_j = \emptyset \qquad \forall i \neq j.$$

Moreover, when the number of voters in the single districts, $|V_j| = v_j$, $j = 1, 2, \ldots, k$, is considered, then $\sum_{j=1}^{k} v_j = V$ must hold.

The Hierarchy of Districts

Many electoral systems adopt a hierarchy of political districts. In this case, the seats are assigned on the basis of successively larger groups of votes obtained by merging districts at higher and higher levels. In other words, a hierarchy of partitions $\mathcal{H} = (\pi_1, \pi_2, \ldots, \pi_p)$ on the set of voters is considered, and each partition π_i is finer than the previous one.

A partition π' is *finer* than a partition π ($\pi' \prec \pi$) if every class of π' is contained in some class of π, as shown in Figure 2.1.

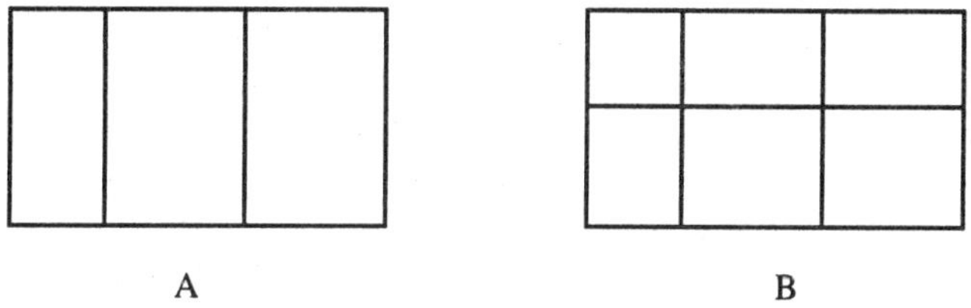

Fig. 2.1. *Partition B is finer than partition A.*

The finest partition ε_D of voters (the one with the largest number of classes) consists of the electoral districts. Each electoral district can be thought of as an "atom," since it cannot be further decomposed. Then let ε_R be a partition obtained by aggregating the classes of ε_D, and let the new (larger) classes be called regions. Finally, the coarsest partition ε_U consists of the single class $\mathbf{V}$, the national district.

Some electoral systems adopt only one level, namely, the basic partition ε_D of electoral districts; others adopt more than one but usually no more than three. The three levels we have listed are such that $\varepsilon_D \prec \varepsilon_R \prec \varepsilon_U$. For technical reasons, we assume that the sequence of partitions always ends with the single class ε_U.

Any such hierarchy of partitions can be represented by a rooted tree. A rooted tree, $T(N, A)$, is a directed graph where all nodes have indegree one, with the exception of a single node of indegree zero, which is the root of the tree, as shown in Figure 2.2. The nodes of the tree represent the classes that form the different partitions at each level. The root corresponds to the single class partition ε_U.

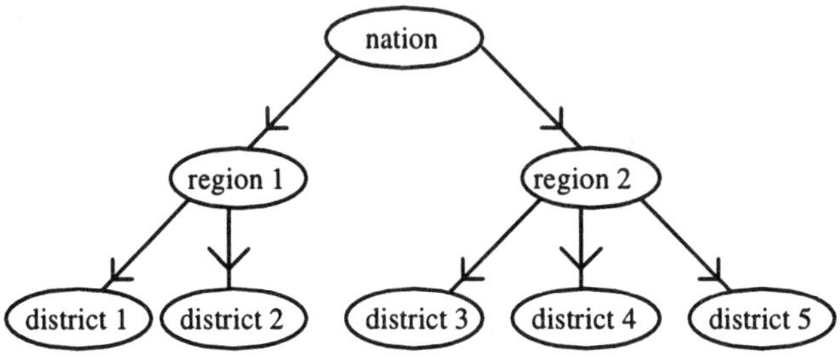

Fig. 2.2. *A rooted tree representing the hierarchy of districts*

Candidate Lineups and Lists

Now consider the set of parties $\mathbf{I} \equiv \{1, 2, \ldots, n\}$ competing in the elections. There is a strict relation between the set of candidates $\mathbf{C}$ and the set of parties $\mathbf{I}$. As a matter of fact, each candidate competes for a unique party, and thus one can define a mapping from $\mathbf{C}$ to $\mathbf{I}$ which associates with each candidate the party he or she runs for. Since each party running in the elections is represented by at least one candidate, the above mapping is surjective.

This kind of relationship is sketched in Figure 2.3, where 10 candidates and three parties are considered:

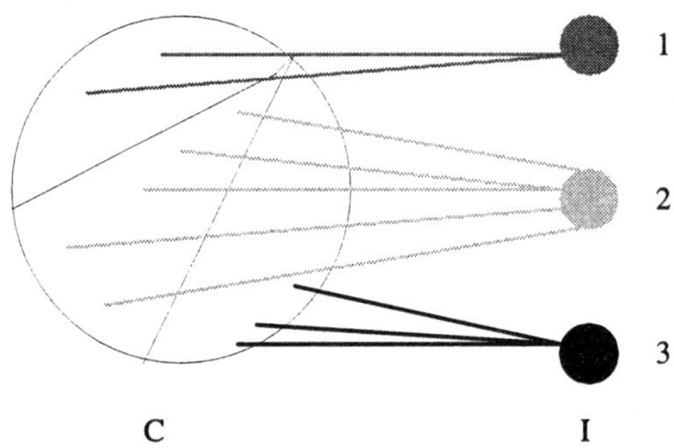

Fig. 2.3. *A surjective mapping from* $\mathbf{C}$ *to* $\mathbf{I}$.

The result of such mapping is nothing more than a partition of the set of candidates in different classes made by the candidates belonging to the same party. Such partition can be denoted by $\mathbf{L} = (L_1, L_2, \ldots, L_n)$, where each subset of candidates L_i is the *lineup* of candidates of party i (Figure 2.4).

As we have already mentioned, the parties partition the set of candidates into different lineups. In fact, what really happens in an election is a bit more complicated: each party selects a (possibly empty) subset of its candidates to present in the single electoral districts in which it will compete. This is an additional operation on the set of candidates belonging to the same party, but it does not necessarily give rise to another partition. In fact, a party can decide

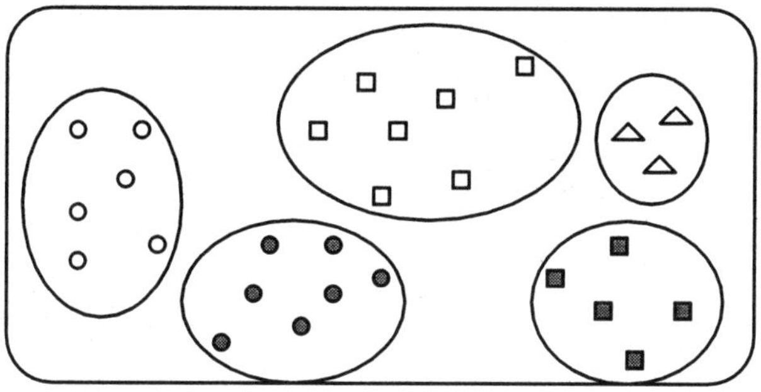

Fig. 2.4. *The set of parties* **I** *induces a partition on the set of candidates* **C**.

to present the same candidate in more than one district (of course, when the districts are single-membered, only one candidate per party can be presented in each district and the same candidate cannot compete in more than one district).

The assignment of candidates to electoral districts can be formally described by a covering $\mathbf{K} = (K_1, K_2, \ldots, K_k)$ of the set $\mathbf{C}$ of candidates, where K_j is the subset of candidates who are running in the jth electoral district. Since every candidate runs at least in one district, one has

$$K_1 \cup K_2 \cup \cdots \cup K_k = \mathbf{C}.$$

However, since a candidate may run in several districts, two subsets K_j and K_l are not necessarily disjoint, but they may partially overlap (see Figure 2.5).

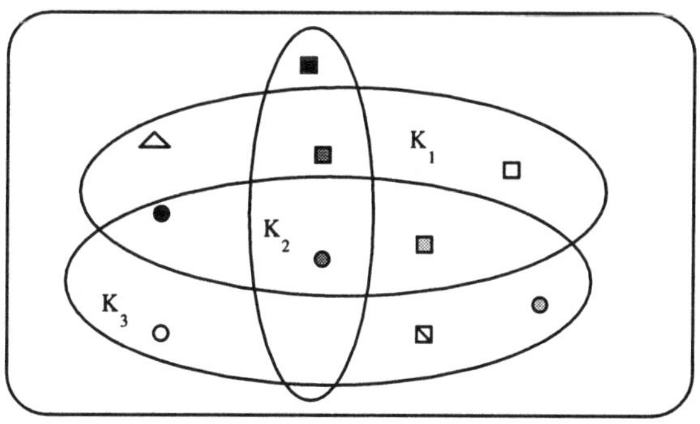

Fig. 2.5. *The three electoral districts define a covering of the sets of candidates; e.g.,* ○ *runs only in district* 3 *while* • *runs in both districts* 1 *and* 3.

From the partition **L** and the covering **K** one can readily obtain, for each party i and for each district j, the subset L_{ij} of those candidates who run for party i in district j: in fact, L_{ij} is just the intersection of L_i and K_j (see Figure 2.6):

$$L_{ij} = L_i \cap K_j.$$

All such subsets of candidates are called *lists*. If $|L_{ij}| = 1$, then L_{ij} is called a *unit list*.

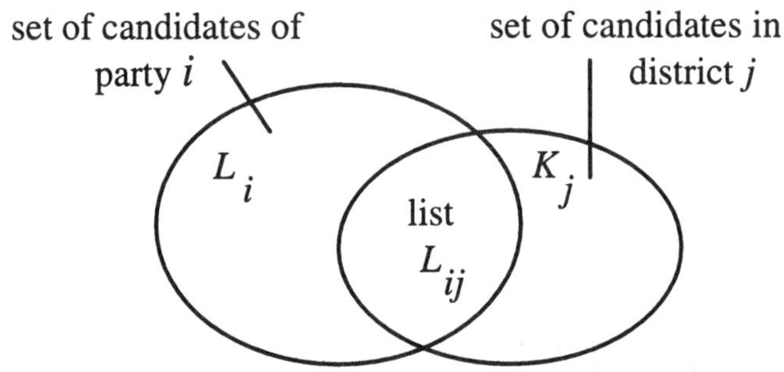

Fig. 2.6. *A list of candidates.*

Party Alliances and Cartels

In several countries, and usually at the higher levels of the district tree T, parties are allowed to form alliances or coalitions for the purpose of seat assignment. In such case, seats are assigned to cartels or groups of lists involving allied parties. Subsequently, all the seats won by each cartel are redistributed among the lists within the cartel. Usually the coalitions do not differ from node to node in T, but there are remarkable exceptions: for example, in the Italian political elections of 1992, "Forza Italia," the party of the media tycoon Silvio Berlusconi, was allied in the northern regions with the separatist Northern League and in the southern ones with the right-wing party, National Alliance. Thus, in order to provide a flexible description of an electoral system, one may define an *alliance mapping* χ, associating with each node q of T a partition π_q of the set $\mathbf{I}$ of parties into *coalitions* $C_1, C_2, \ldots, C_m$ of parties. For convenience, let the n parties be labelled $i = 1, 2 \ldots, n$. Each coalition C_h is a subset of $\{1, 2, \ldots, n\}$. As mentioned above, node q represents a set of districts: let $J_q \subseteq \{l, \ldots, k\}$ be the set of indexes of these districts and let $\mathcal{L}_q \equiv \{L_{ij} : i = 1, 2, \ldots, n; j \in J_q\}$ be the set of all lists associated with node q. The set of lists

$$L_q(C_h) = \{L_{ij} : i \in C_h, j \in J_q\}, \quad h = 1, 2, \ldots, m,$$

are the *cartels* associated with the partition π_q. There is one cartel for each coalition of parties, and altogether the m cartels form a partition of $\mathcal{L}_q$ (see Figure 2.7, where $n = 8$, $k = 9$, $m = 3$).

Therefore, a cartel is a cross section of the set $\mathcal{L}$ of all lists and is completely specified by

(1) a subset C_h of parties;

(2) a subset J_q of districts.

In order to provide a uniform description of the electoral engine, we shall make the assumption that, at the bottom level (district level) of the tree T, every cartel consists of a single list.

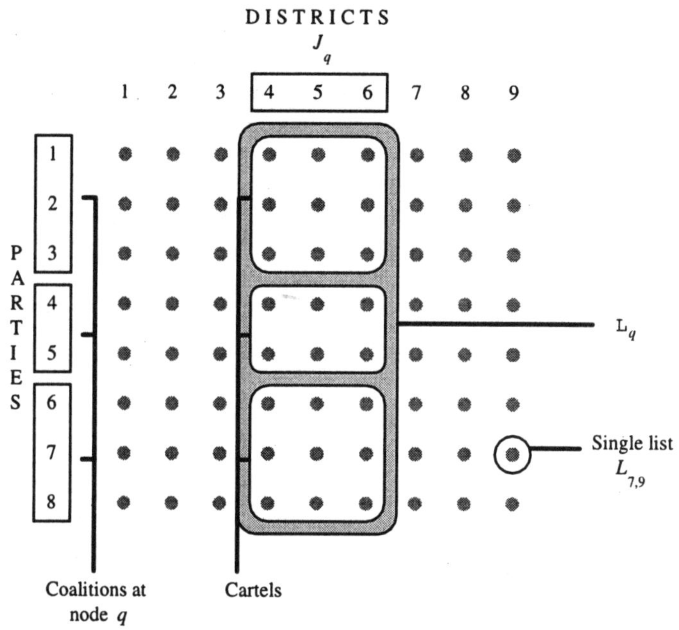

Fig. 2.7. *Coalitions and corresponding cartels.*

Seat Apportionment

The set **S** of seats to be assigned as a result of the vote is another basic element of the system. In some countries, seats are assigned to parties and eventually to candidates, altogether at the national level. More often, seats are preliminarily allotted to the individual districts, with the provision that every seat assigned to a given district must go to some candidate running for that district. In some other cases, a fraction of seats is assigned to the districts and the remaining seats are allotted to groups of districts (regions or the whole nation).

In order to obtain a general definition of seat apportionment, recall the notion of a hierarchy of districts and its representation as a rooted tree $T(N, A)$. Each node of T represents a class of some partition in the hierarchy, that is, a district, a region, the nation. In general, a *seat apportionment* may be defined as a mapping α from a set **S** of seats to the node set N of T: the mapping may assign a seat either to a district, or to a region, or to the whole nation. Of course, some nodes may get no seats. The underlying rule is that every seat attributed to a district may be won only by a candidate competing there; if a seat is allocated to a region, the winner must run in at least one of the districts of the region.

The districts are *single-membered* if only one seat is associated with the corresponding territorial district, and they are *multimembered* if there are more seats associated with the same territorial district. The number of seats associated with the electoral district is defined by the *size* (or *magnitude*) of the district. In case of multimember districts we talk about *small* or *big* districts according to the number of seats in that district, which can vary from two and three to, say, a hundred.

It should be emphasized that, while the actual vote takes place, as a rule, only at the district level, the assignment of seats to parties and to their candidates may occur also at other levels of the tree.

Moreover, in some countries, such as Germany, Austria, Denmark, Iceland, and Sweden, supplementary seats may be granted on the basis of the vote outcome.

Ballot Structure

During the election, each voter casts a *ballot*. The information contained in the voters' ballots is processed by the electoral formula (including tie-breaking rules, etc.) to determine the number of seats assigned to the different parties and also the actual assignment of these seats to elected candidates.

The ballot structure is an essential component of an electoral system and it may take several alternative forms. According to Rae (1967) a ballot is *categorical* when "the voter gives his mandate to [one or more candidates of] a single party" and *ordinal* when the voter "divides his [mandate] among parties or among candidates of different parties."

The functional viewpoint we are adopting here calls for a finer classification. With reference to a given district j, we classify ballots into four different groups:

- *list ballots*, containing a single list L_{ij};
- *preference ballots*, containing a subset of at most r candidates in K_j, where the integer r is fixed in advance and is called the *number of preferences*;
- *cumulated ballots*, containing a bag (subset with possibly repeated elements) of candidates in K_j. Each elector is granted a "stock" of at most r preferences, which may be arbitrarily distributed among no more than r candidates in K_j. Candidates are allowed to receive multiple preferences, even the full stock;
- *ranking ballots*, containing a full ranking (i.e., a permutation) of K_j in decreasing order of preference.

Preference ballots are further categorized with respect both to the number of preferences and to the possibility of *panachage* (voting for candidates of different parties) as follows:

- *single-preference ballots*, which contain just one candidate in K_j;
- *multiple-preference ballots*, which contain at most r candidates in K_j. If it is required that all the chosen candidates belong to the same party then the ballot is called *homogeneous*; otherwise (when *panachage* is permitted) the ballot is called *heterogeneous*.

The full classification tree is shown in Figure 2.8, where the asterisks (*) stand for ordinal ballots according to Rae's terminology, while the remaining ones are categorical.

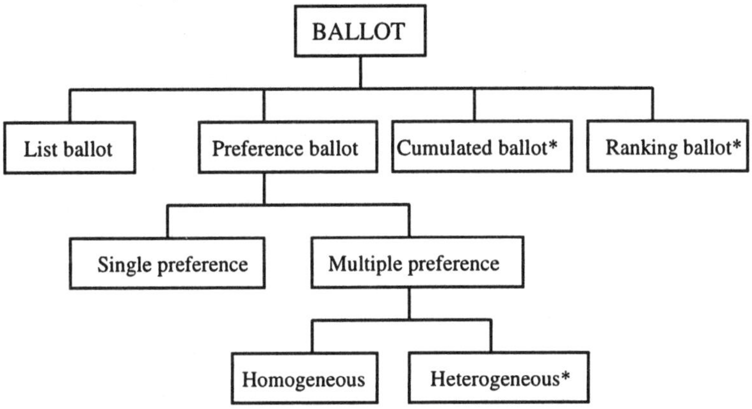

Fig. 2.8. *Classification of the possible ballot types.*

The set of all possible ballots in district j will be denoted by Ω_j. We denote by Ω the k-tuple $(\Omega_1, \Omega_2, \ldots, \Omega_k)$. If the elements of Ω_j are list ballots, then $|\Omega_j| = l_j$ is the number of lists in district j. If they are preference ballots, then $|\Omega_j|$ is equal to the *binomial coefficient* $\binom{q_j}{r} = \frac{q_j!}{r!(q_j-r)!}$, where $q_j = |K_j|$ and, as usual, $m!$ denotes the *factorial* $m \times (m-1) \times (m-2) \times \cdots \times 3 \times 2 \times 1$. If they are cumulated ballots, then $|\Omega_j| = \binom{q_j+r-1}{r}$; to make things simpler, for the latter two types of ballot we have assumed that each voter takes *exactly* r preferences. Finally, if they are ranking ballots, then $|\Omega_j| = q_j!$.

Single or Multiple Ballots

The next feature we want to analyze is the number of ballots. In fact, some systems (in particular systems using single-membered districts) expect the voters to vote more than once to select the winning candidates. The idea underlying such methods is that the final selection can be made by successive elimination of the less voted candidates. In other words, the first ballot is used to cancel a set of candidates so that at the next ballot a smaller set is considered, just as if several subelections are performed each time, until only one candidate is left.

Methods that adopt more than one successive ballot will rank the candidates into three categories on the basis of the partial result obtained each time. In fact, after each ballot the candidates can be distinguished into elected candidates, candidates who are deferred to the next ballot, and candidates who are definitively eliminated from the election. Given the set of candidates $\mathbf{C}$, the balloting B may be described by a function $B : \mathbf{C} \to \{-1, 0, 1\}$, while the number of subelections the method can carry out is denoted by t. The function assigns -1 to a candidate who will be canceled from the set, 0 to a candidate who remains, and 1 to a candidate who is elected.

Vote and Vote Counts

In every district j, each elector expresses his or her preferences for the lists L_{ij} and possibly for the candidates in K_j by choosing a specific ballot in the set Ω_j of all possible ballots. Thus within district j the vote may be formally described as a mapping φ_j of the set V_j of electors in district j into the set Ω_j.

Globally, the vote is a k-tuple $\varphi = (\varphi_1, \varphi_2, \ldots, \varphi_k)$ of mappings, one for each district. On the other hand, the principles of democracy and the reasons of anonymity require that the electors be *indistinguishable*. Hence, what matters in practice is not the individual vote cast by each elector, but the *vote count*, which summarizes the outcome of the individual votes and constitutes the input for the electoral engine, whose final output is the assignment of seats to candidates. Actually, in the case of list ballots the vote count is defined by the number v_{ij} of votes obtained by each list L_{ij}; in the case of preference or cumulated ballots, the vote count is defined by the number π_{hj} of preferences (possibly with their multiplicities) obtained by each candidate h; in the case of ranking ballots the vote count is the number $\omega_{\rho j}$ of votes obtained by each ranking (permutation) ρ of the set K_j.

Seat Assignment

The basic task of an electoral system is the assignment of seats to elected candidates. Notice that what matters is not only *who* is elected but also *which seat* goes to each elected candidate. As a matter of fact seats are usually apportioned to specific districts or to specific regions, and thus they have strong territorial features. On this basis, one may formally define a seat assignment as follows. Let $B : C \to \{-1,\ 0,\ 1\}$ be the balloting function relative to the ballot under consideration, and let

$$E = \{c \in C : B(c) = 1\}$$

be the set of new elected candidates.

Then a *seat assignment* is defined as a mapping σ from E to S. This mapping specifies, for each elected candidate c, the actual seat $\sigma(c)$ assigned to c. Since different candidates get different seats, the mapping σ is *injective* (see Figure 2.9). On the other hand, σ might not be *surjective*, since after the current ballot there might still be some unassigned seats.

Electoral Engine

Once the electors have cast their votes, these are fed into a mechanism whose eventual output is the assignment of seats to candidates. In all electoral rounds but the final one, a decision must also be made about which candidates should be admitted to the next round. Formally, for each round there is an electoral

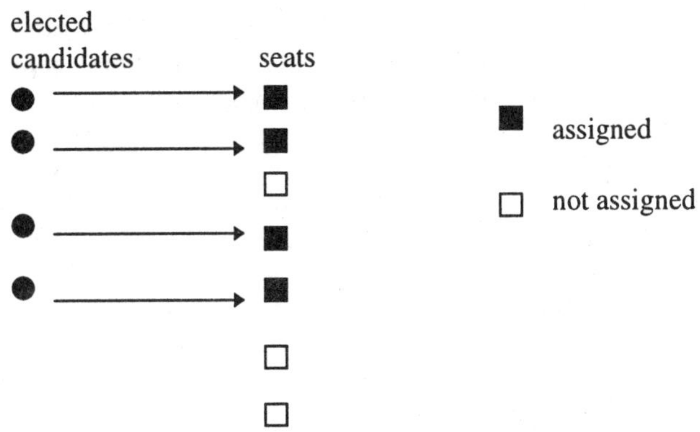

Fig. 2.9. *Injective mapping from the set of elected candidates to the seats.*

engine associating with each possible vote φ both a balloting function B and a seat assignment σ. Such a mechanism performs a twofold task:

(1) *conversion* of votes into seats assigned to the different cartels;

(2) *selection* within each cartel of a subset of elected candidates, whose number matches the number of seats won by that cartel, as well as a subset of candidates promoted to the next ballot, if any.

Sometimes the above two tasks are performed simultaneously. A simple example is given by those systems where unit lists are adopted: if a seat is assigned to one such list as a result of the vote, then the unique member of the list is automatically elected and gets that seat. Another example is provided by systems where ranking ballots are adopted. Here the electoral engine directly yields a set of elected candidates. This set induces an assignment of seats to parties, since each candidate runs for exactly one party.

In other cases, the two tasks are performed sequentially. Consider, for instance, systems where lists are allowed to contain several candidates. Such lists are preliminarily ordered by the corresponding parties in such a way that if a list gets s seats then the first s candidates in the list are selected (in this case one speaks of *blocked* lists). Another notable example is given by systems involving preference ballots (or cumulated ones). The electoral engine here again exploits the unique association of each candidate with a party: each ballot where a preference is expressed in favor of one or more candidates belonging to a certain list is counted as one ballot in favor of that list. After list votes are converted into list seats, the preference ballots are used again to select within each list the most voted candidates.

Since in general candidates may compete in several districts, it is quite possible that some candidate turns out to win in more than one district. Hence, the electoral engine must embody additional rules that make multiple winners cover exactly one seat each and then assign the remaining seats to further candidates belonging to the same party.

Let us examine more closely how the mechanism works, after each ballot, at an arbitrary node q of the district tree T. Recall that q represents either a district, or a region, or the whole nation (when q is the root of T). With reference to Figure 2.10, there is an inflow u_q of votes coming either directly from the electors (this happens when q is a leaf of the tree, i.e., q represents a district) or from a previous level of the tree. The votes are attributed either to lists or to candidates or to rankings, depending on the ballot structure, as discussed above. Then a *filter* selects the *active* votes that are actually used to assign seats at the level of hierarchy under consideration. In most cases, the active votes coincide with the total inflow u_q, but in some cases, the active votes represent only a part of the total inflow (we refer, in particular, to the detraction rules used in many mixed or hybrid systems as further described in Chapter 8).

Subsequently, a *converter* device transforms votes attributed to lists into seats assigned to cartels (this is what is usually referred to as electoral formula). The total number of assigned seats must be equal to the number S_q of seats apportioned to node q.

The filter also outputs an outflow v_q of votes going to the next (higher) level, namely, into the *father* of node q (the only node of T from which there is an arc going into q). Since the active votes might be "burnt" (because of detractions, exclusion thresholds, and other special devices) in the vote-to-seat conversion, one has $v_q \leq u_q$.

The seats assigned to each cartel are then redistributed by a *distributor* among the lists within the cartel.

Finally, taking into account the number of seats assigned by the electoral formula to the different lists and, possibly, the information about the voters' preferences for candidates, the last device (the *selector*) partitions the set C_q of the candidates associated with node q into three subsets:

- E_q, the subset of elected candidates, where $|E_q| = S_q$;
- D_q, the subset of candidates who are deferred to the next ballot;
- R_q, the subset of rejected candidates.

Putting Everything Together: The General Electoral System

After having discussed all the basic ingredients of an electoral system, we are back to the initial question: what is an electoral system? On the basis of the above discussion, we propose the following definition: an *electoral module* is a 10-tuple

$$S = (\mathbf{V}, \mathbf{C}, \mathbf{I}, \mathbf{S}; \mathcal{H}, \alpha; \mathcal{L}; \chi, \Omega; \mathcal{M}),$$

where

- $\mathbf{V}$ is the set of voters;

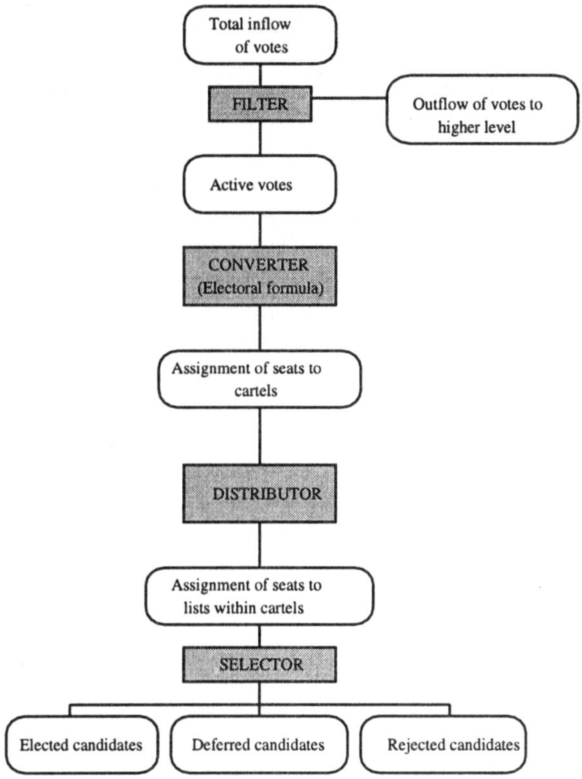

Fig. 2.10. *Electoral engine in a node of the district tree.*

- **C** is the set of candidates;
- **I** is the set of parties;
- **S** is the set of seats;
- $\mathcal{H} = (\pi_1, \pi_2, \ldots, \pi_p)$ is a hierarchy of districts, represented by a rooted tree $T(N, A)$;
- $\alpha : \mathbf{S} \to N$ is the seat apportionment;
- $\mathcal{L} = \{L_{ij} : i = 1, 2, \ldots, n;\ j = 1, 2, \ldots, k\}$ is the set of lists;
- χ is the alliance mapping from N to the set of partitions of **I**;
- $\Omega = (\Omega_1, \Omega_2, \ldots, \Omega_k)$ is a k-tuple where each element denotes the set of possible ballots adopted in a district;
- $\mathcal{M}$ is the electoral engine, mapping each possible vote φ into a balloting function D and a seat assignment υ.

In the following sections examples of the main electoral systems are provided in further detail.

An *electoral system* is a sequence of electoral modules $\boldsymbol{S}^{(1)}, \boldsymbol{S}^{(2)}, \ldots, \boldsymbol{S}^{(t)}$, where t is the total number of ballots, such that only the last module $\boldsymbol{S}^{(t)}$ does not allow for deferred candidates.

Since for most of the examples given in the following pages the vote-to-seat transformation is independent from the specific district under consideration, for the sake of readability, we will drop the suffix j when it is superfluous. However, in case such notation may be misleading, the alphabet adopted is changed.

Plurality Systems

Among the electoral systems currently adopted in democratic countries, the majority systems are, in fact, a minority. Nevertheless, the first-past-the-post system is largely used in Anglo-Saxon countries, such as Great Britain, the United States, and New Zealand, as well as in Italy; the double ballot is adopted in France and was also used in Italy between 1861 and 1919 (except for a short period 1882–1892); the alternative transferable vote is adopted in Australia.

First-Past-the-Post

The philosophy of this system is "winner takes all" and nothing is left to the losers. It is the most drastic electoral system, and perhaps the most ancient, based on the principle of territorial representation. In fact, the territory is divided into a number of districts equal to the number of candidates that must be elected. In every district, each party presents only one candidate. The most voted candidate wins the seat independently from the percentage of votes he has obtained. The first-past-the-post can thus be specified by

- the four sets **V**, **C**, **I**, **S** of voters, candidates, parties, and seats, respectively;
- $k = S$ districts $V_1, V_2, \ldots, V_k$;
- $\mathbf{K} = \{K_1, K_2, \ldots, K_k\}, |K_i| \leq n, \; i = 1, 2, \ldots, k;$[1]
- the apportionment mapping α allocates one seat to each district;
- all $L_{ij} \in \mathcal{L}$ are unit lists;
- since list ballots are used, Ω_j is the set of all nonempty lists L_{ij}, $i = 1, 2, \ldots, n$.

Consider a district j. Let v_{ij} be the number of votes obtained by the unit list L_{ij}, and let $M_j = \max_{1 \leq h \leq n} v_{hj}$ be the maximum number of votes obtained by some list, say, L_{ij}. In the vast majority of cases, the maximum M_j of votes is attained by a unique list L_{ij}. Should this not be the case, tie-breaking rules

[1]The equality is verified when *each* party presents only one candidate in each district.

are invoked, or the election is repeated. A similar remark applies also to other majoritarian systems.

Then the unique seat allocated to district j goes to list L_{ij} and the mapping σ assigns this seat to the only candidate in L_{ij}.

Double Ballot

This method, also known as the French system, is articulated into two electoral rounds completed in a brief lapse of time. During the first round, seats are assigned only to those candidates who have received an absolute majority of votes, that is, over 50% of the total preferences in a district. For districts in which this does not happen, a second round is needed. Usually, only the candidates who have gained a certain threshold of votes τ are promoted to the second round, where a seat is assigned, in each district, to the candidate who simply receives the biggest vote share.

The second round can be *open*, that is, all candidates can participate in the second round; *partially closed*, when candidates are admitted if they obtained a fixed threshold of votes (for example, in France 12.5% of the votes); *totally closed*, if only the first two candidates remain in the competition (as in Italy for the election of the mayor).

The *double-ballot method* is defined by

- the four sets **V**, **C**, **I**, **S** of voters, candidates, parties, and seats, respectively;
- $K = S$ districts $V_1, V_2, \ldots, V_k$;
- $\boldsymbol{K} = \{K_1, K_2, \ldots, K_k\}$, $|K_i| \leq n$, $i = 1, 2, \ldots, k$;
- the apportionment mapping α allocates one seat to each district;
- all $L_{ij} \in \mathcal{L}$ are unit lists;
- since list ballots are used, Ω_j is the set of all nonempty lists L_{ij}, $i = 1, 2, \ldots, n$,

just as in the first-past-the-post system.

Consider district j. Let s_j be the only seat allocated to such a district. Let v_j be the total number of votes registered there, and let v_{ij}^1 be the number of votes received by list L_{ij}—and by its only candidate c_{ij}—in the first round. Let

$$M_j = \max\{v_{ij}^1 : 1 = 1, 2, \ldots, n\}.$$

Then the balloting function B is defined as follows:

$$B(c_{ij}) = \begin{cases} 1 & \text{if } v_{ij}^1 = M_j \text{ and } M_j > \frac{1}{2}v_j, \\ 0 & \text{if } v_{ij}^1 = M_j \text{ and } \tau \leq M_j \leq \frac{1}{2}v_j, \\ -1 & \text{otherwise,} \end{cases}$$

where $\tau(0 \leq \tau \leq \frac{1}{2})$ is a lower threshold, defining the minimum percentage of votes whereby candidates are promoted to the second round. When the second round is open $\tau = 0$; when it is partially closed τ is a fixed percentage; when it is totally closed, a variable threshold can be considered from district to district, corresponding to the percentage of votes obtained by the second-best list.

Let $J \equiv \{j$: there is some candidate i such that $B(c_{ij}) = 0\}$. In other words, J is the index set of those districts where no candidate was elected after the first round.

In the second round, the sets of voters, candidates, parties, and seats are given by

$$\begin{aligned} V' &\equiv \bigcup_{j \in J} V_j, \\ C' &\equiv \{c_{ij} : B(c_{ij}) = 0\}, \\ I' &\equiv \{i : B(c_{ij}) = 0 \text{ for some district } j\}, \\ S' &\equiv \{s_j : j \in J\}, \end{aligned}$$

respectively. The second electoral round follows the rules of a first-past-the-post system.

Alternative Transferable Vote

Within a single-member district, each elector has a single vote and the possibility of expressing preferences for other candidates. The seat is given to a candidate when he or she reaches the absolute majority of votes. If nobody obtains it, the candidate who has obtained the minimum number of votes is eliminated and his votes are shared among the other candidates according to the second preferences expressed. This elimination procedure continues until one candidate obtains more than 50% of votes.

In this case the parameters are defined as follows:

- the four sets $\mathbf{V}$, $\mathbf{C}$, $\mathbf{I}$, $\mathbf{S}$ of voters, candidates, parties, and seats, respectively;
- $K = S$ districts $V_1, V_2, \ldots, V_k$;
- $\boldsymbol{K} = \{K_1, K_2, \ldots, K_k\}, |K_i| \leq n, \ i = 1, 2, \ldots, k$;
- the apportionment mapping α allocates one seat to each district;
- ranking ballots are used, and thus Ω_j is the set of all permutations of K_j.

Again, consider district j, and let v_j be the number of its voters. The initial set of eligible candidates is $K_j^{(1)} = K_j$. Let $v_{hj}^{(1)}$ be the number of ballots where candidate h ranks first for all $h \in K_j^{(1)}$.

If there exists a candidate $a \in K_j^{(1)}$ such that $v_{aj}^{(1)} > \frac{1}{2}v_j$, then a is elected and gets the seat allocated to district j. Otherwise, let b be the candidate in $K_j^{(1)}$ such that

$$v_{bj}^{(1)} = \min\{v_{hj}^{(1)} : h \in K_j^{(1)}\}.$$

Now, the set of eligible candidates becomes $K_j^{(2)} = K_j^{(1)} \setminus \{b\}$.

For all h in $K_j^{(2)}$, let $v_{hj}^{(2)}$ be the number of ballots where h ranks first among all candidates in $K_j^{(2)}$. In other words, $v_{hj}^{(2)}$ is equal to the number of ballots where h ranks first, plus the number of those where b ranks first and h ranks second.

If there exists a candidate $a \in K_j^{(2)}$ such that $v_{aj}^{(2)} > \frac{1}{2}v_j$, then a is elected and gets the seat allocated to district j. Otherwise, let c be the candidate in $K_j^{(2)}$ such that

$$v_{cj}^{(2)} = \min\{v_{hj}^{(2)} : h \in K_j^{(2)}\}.$$

Then the set of eligible candidates becomes $K_j^{(3)} = K_j^{(2)} - \{c\}$; for all $h \in K_j^{(3)}$, let $v_{hj}^{(3)}$ be the number of ballots where h ranks first among all candidates in $K_j^{(3)}$. In other words, $v_{hj}^{(3)}$ is the number of ballots whose rankings begin as follows:

1°	2°	3°	4°
h	-	-	-
b	h	-	-
c	h	-	-
b	c	h	-
c	b	h	-

The process is repeated until, at a certain iteration s, one finds a candidate $a \in K_j^{(s)}$ such that $v_{aj}^{(s)} > \frac{1}{2}v_j$. This must necessarily occur sooner or later: in the worst case, when $K_j^{(s)}$ contains only one candidate, this candidate must get all the votes!

Proportional Systems

Most countries adopt proportional electoral formulas that are designed to reduce, as far as possible, the distortion created by transforming the set of votes into a much smaller set of seats. The proportional formulas are generally classified into two groups: the quotient methods and the divisor methods. Both, however, contain many variants, about 11 of which are currently used in some European countries, such as Austria, Spain, Portugal, Belgium, and all Scandinavian countries.

Quotient (or quota) methods are based on the ratio between the number of votes obtained and the number of seats to assign. On the basis of the integer part of this *quotient* they proceed to distribute seats among cartels. This procedure yields some remainders, that is, a certain number of seats that cannot directly be assigned. Remainders are distributed on the basis of different rules but more commonly on the basis of the Largest Remainders. Variants based on the definition of the quotients are the natural quota method (also called largest remainders), the Droop (or +1) method, and the Imperiali (or +2) method.

Divisor methods are instead based upon a set of *divisors*. For each cartel and each divisor, one computes the ratio between the number of votes received by that cartel and the divisor considered. Then the S seats are assigned to the cartels achieving the S largest such ratios. Of course, in this way a cartel may obtain several seats.

These methods do not produce remainders. The most common divisor methods are d'Hondt, Sainte-Laguë, modified Sainte-Laguë, the Belgian method, the Danish method, Equal Proportions, and Smallest Divisors.

We assume that list ballots are used: hence $\Omega_j = \{L_{ij} : i = 1, 2, \ldots, n\}$ for every district j. In general, consider any given node q of the tree T, and its cartels

$$H_1 = L_q(c_1), \ldots, H_m = L_q(c_m).$$

For each cartel H_i, let v_i be the total number of votes obtained by H_i (that is, the sum of the votes received by all lists within H_i; without loss of generality, we shall assume that $v_i > 0$), and let s_i be the number of seats to be assigned to H_i.

Furthermore, let $\mathcal{P} = v_1 + \cdots + v_m$ be the total number of votes at node q and $\mathcal{S}$ the total number of seats apportioned to this node. Ideally, for the sake of equity, one would like to have $\frac{s_1}{v_1} = \cdots \frac{s_m}{v_m} = \rho$ for some positive constant ρ. If this is the case, one must have $\rho = \frac{\mathcal{S}}{\mathcal{P}}$. In fact, by summing up the equalities $s_i = \rho v_i$, $i = 1, 2, \ldots, m$, one gets $\mathcal{S} = \rho\mathcal{P}$. Thus, if all ratios $\frac{s_i}{v_i}$ were equal, the number of seats assigned to cartel H_i would be given by $s_i = \frac{\mathcal{S}}{\mathcal{P}} v_i$.

Unfortunately, the numbers s_i defined by the above formula are generally not integers. All proportional formulas aim to determine integral values of $s_1, \ldots, s_m$ in such a way that the ratios $\frac{s_i}{v_i}$ are approximately equal to each other and to the ratio $\frac{\mathcal{S}}{\mathcal{P}}$.

Quotient Methods

The common idea of these methods is to assign to each cartel H_i a number s_i of seats obtained by rounding either up or down a given quota. Such quota is equal to $\frac{\mathcal{S}}{\mathcal{P}} v_i$ in the case of the natural quota, to $\frac{\mathcal{S}+1}{\mathcal{P}} v_i$ in the case of the Droop quota, and to $\frac{\mathcal{S}+2}{\mathcal{P}} v_i$ in the case of the Imperiali quota.

Let us consider the case of the natural quota. For all $i = 1, 2, \ldots, m$ one gets

$$\lfloor q_i \rfloor \leq s_i \leq \lceil q_i \rceil,$$

where $\lfloor z \rfloor$ and $\lceil z \rceil$ and are the largest integer not exceeding z and the smallest integer greater than or equal to z, respectively. If z is an integer, then $\lfloor z \rfloor = \lceil z \rceil$; if not, $\lceil z \rceil = \lfloor z \rfloor + 1$.

Thus, for each cartel H_i one must face a binary choice: should H_i get only $\lfloor q_i \rfloor$ seats or one more? One can build different methods by performing this choice in a different way. Nevertheless, the popular way to do this is the *largest remainders* method: for each $i = 1, 2, \ldots, m$ compute the *remainder*:

$$r_i = q_i - \lfloor q_i \rfloor.$$

The total number of cartels H_i that get one extra seat is then given by $R = \mathcal{S} - \sum_{i=1}^{m} \lfloor q_i \rfloor$. These R extra seats are assigned to the cartel having the R largest remainders.

The same kind of rule can be used with the Droop and Imperiali quota, too. However, the natural quota method with Largest Remainders is usually (and henceforth) the one referred to by Largest Remainders for short.

Divisor Methods

All methods in this class are based on a *divisor criterion*: by this we mean a function $d(s)$ defined for all natural numbers $s = 0, 1, 2, \ldots$ and taking values in the set $\Re$ of real numbers,[2] such that

$$0 \leq d(0) \leq 1 \leq d(1) \leq 2 \leq \cdots$$

and

$$d(0) < d(1) < d(2) < \cdots.$$

Then, for each $i = 1, 2, \ldots, m$ and each $s = 0, 1, \ldots, \mathcal{S} - 1$ the ratios

$$\rho_{is} = \frac{v_i}{d(s)}$$

are computed and the $\mathcal{S}$ largest ratios are selected (ties being arbitrarily broken). For each selected ratio ρ_{is}, one seat is assigned to cartel H_i.

In some divisor methods (such as Adams's), $d(0)$ is taken to be zero. In such case, one makes the convention that $\frac{v_i}{0} = +\infty$, where the "plus infinity" symbol $+\infty$ is such that $\min\{+\infty, a\} = a$ and $\max\{+\infty, a\} = +\infty$ for every given real number a. This implies that each party gets at least one seat.

Table 2.1 shows the main divisor methods in use, the sequence of divisors they use, and their corresponding divisor criterion. In Table 2.2 examples of largest remainder methods are provided.

Single Transferable Vote

The single transferable vote (STV) allocation method has many similarities with the alternative transferable vote described above. Both methods use ranking ballots and both can be described as iterative procedures where the candidate with the smallest current number of votes is eliminated, his votes being then

[2] Loosely speaking, a *real number* is any number that in decimal notation can be represented by a finite sequence of digits (possibly starting with a minus sign), followed by a decimal point and by another (finite or infinite) sequence of digits. All digits are integers between 0 and 9, and the decimal point may be omitted if it is not followed by any digit. For example, the following are real numbers: $12, 12.25, -12.25, 0.12333\ldots$, $\pi = 3.14159265\ldots$.

Table 2.1. *Examples of divisor methods.*

Method	*nth divisor*	*Sequence of divisors*	*Divisor criterion*	*Examples*
Belgian	$(n+1)/2$	$1, 1.5, 2, 2.5, 3, \dots$		Belgium
d'Hondt	n	$1, 2, 3, 4, 5, \dots$	$s+1$	Austria, Belgium, Finland, Island, Spain, Portugal, Holland, Switzerland, France (1986)
Sainte-Laguë	$2n-1$	$1, 3, 5, 7, 9, \dots$	$s+\frac{1}{2}$	Denmark (1945–1953)
Sainte-Laguë modified	$(10n-5)/7$	$1, 2.14, 3.57, 5, 6.43, \dots$		Norway, Sweden
Equal Proportions	$n(n-1)$	$0, 1.41, 2.45, 3.46, 4.47, \dots$	$\overline{s(s+1)}$	USA
Danish	$(3n-2)$	$1, 4, 7, 10, 13, \dots$		Denmark
Smallest Divisors	$(n-1)$	$0, 1, 2, 3, 4, \dots$	s	-
Nohlen	$(n+1)$	$2, 3, 4, 5, 6, \dots$		

transferred to his next most preferred candidates who are still in competition. However, besides this "bottom-up" transfer of votes, in the STV a "top-down" vote transfer also takes place: the candidates whose current number of votes exceeds a certain threshold are definitively elected, and their votes in excess of the threshold transferred to their most preferred noneliminated candidates by a proportional allocation rule.

Another basic difference with the alternative transferable vote system is that in the latter method, as mentioned above, all districts are single membered, whereas in STV they may be also multimembered.

A formal description of the STV method follows. There are k districts and the apportionment mapping α allocates $S_1, S_2, \dots, S_k$ seats to districts $1, 2, \dots, k$, respectively, where $S_1 + S_2 + \cdots + S_k = S$. In each district j, since ranking ballots are used, Ω_j is the set of all permutations of the set of candidates running in that district.

Consider any given district. Let $\mathcal{S}, \mathcal{P}, \mathcal{K}$ be the number of seats allocated to that district, the number of voters within the district, and the number of its candidates, respectively. The minimum number of votes any candidate must get in order to be elected is the so-called *Droop quota* (see Chapter 5 for an explanation):

$$\vartheta = \left\lfloor \frac{\mathcal{P}}{\mathcal{S}+1} \right\rfloor + 1 .$$

Initially, the set of eligible candidates is $K^{(1)} = \mathcal{K}$, and each candidate h is attributed a number $v_h^{(1)}$ of votes given by the number of ballots where h ranks

Table 2.2. *Examples of largest remainders methods.*

Method	*Quota*	*Examples*
Natural quotient (or Largest Remainders)	$\frac{P}{S}$	Austria,* Belgium,* Denmark,* Germany
Droop quota	$\frac{P}{S+1}$	Greece
Imperiali quota	$\frac{P}{S+2}$	Italy (until 1992)

* Some countries use different proportional methods at different levels in the hierarchy of districts.

first. Let us describe a generic iteration t of the method.

Let $K^{(t)}$ be the set of those candidates who are still in competition at the beginning of the tth iteration, that is, those who have not been elected or eliminated yet. Let $A^{(t)}$ be the set of those candidates $h \in K^{(t)}$ whose current number of votes $v_h^{(t)}$ is at least ϑ. We distinguish two cases.

Case 1. $A^{(t)}$ *is nonempty*. Let $v_a^{(t)} = \max_{h \in A^{(t)}} v_h^{(t)}$. Then candidate a, who holds the largest number of current votes, is elected and gets one seat. Let $R^{(t)} = (v_a^{(t)} - \vartheta)$ be the number of *excess votes* of candidate a. Now the set of eligible candidates becomes $K^{(t+1)} = K^{(t)} - \{a\}$. Among the ballots where candidate $h \in K^{(t+1)}$ ranks first in $K^{(t+1)}$, let $u_h^{(t)}$ and $w_h^{(t)}$ be the number of ballots where h is preceded or followed by a, respectively. Allocate $R^{(t)}$ excess votes of a to the candidates of $K^{(t+1)}$ as proportionally as possible to their "scores" $u_h^{(t)}$ with the method of largest remainders: let $r_h^{(t)}$ be the number of votes attributed to $h \in K^{(t+1)}$ in this way. Finally, set $v_h^{(t+1)} = w_h^{(t)} + r_h^{(t)}$ for all $h \in K^{(t+1)}$, and go to iteration $t+1$.

Case 2. $A^{(t)}$ *is empty*. In this case the same mechanism as the alternative transferable vote is used. Let c be the candidate in $K^{(t)}$ such that $v_c^{(t)} = \min_{h \in K^{(t)}} v_h^{(t)}$. Then candidate c is eliminated, the new set of candidates becomes $K^{(t+1)} = K^{(t)} - \{c\}$, and the new number of votes $v_h^{(t+1)}$ attributed to each candidate h in $K^{(t+1)}$ is given by the number of ballots where h ranks first among all candidates in $K^{(t+1)}$. Notice that $v_h^{(t+1)}$ is equal to the previous number of votes $v_h^{(t)}$, plus the number of votes *transferred* from c to h: the latter number is precisely equal to the number of ballots where h ranks first among the candidates in $K^{(t+1)}$, but he or she is preceded by candidate c. Go to iteration $t+1$.

Such iterative procedure stops when all the S seats have been assigned.

The STV is traditionally considered to be a proportional method for its use of the Droop quota and for its proportional redistribution of excess votes from elected candidates to second choice ones. It is used, with minor variations, in Ireland and Malta.

2.2 Coding Electoral Systems

As can be noticed from the previous section, a functional description of elementary electoral systems can be very complicated. In the examples above, only the basic features of each system are considered, but in real applications, electoral systems usually include a great number of refinements such as thresholds, detractions, majority bonuses, etc., which would be impossible to incorporate without generating unreadable notation.

We suggest characterizing different classes of electoral systems by using a suitable coding system (just as is done in scheduling [Graham, Lawler, Lenstra, and Rinnooy Kan, 1979]). Representing each system by a particular string of variables, we can easily produce a complete classification of the huge number of existing systems without having to refer to complicated explanations and tedious repetitions in the text. Moreover, "coding" allows us to generate all possible electoral systems through the feasible combinations of these variables in a simple and effortless manner.

The elements that characterize an electoral system are essentially three:

- type of ballot (α);
- hierarchy of districts, cartels, and seat apportionment (β);
- type of formula (γ).

Each system is identified by a particular triple $\alpha|\beta|\gamma$ of strings, where $\alpha \in \{$*list, pref, cum, rank*$\}$, that is, list ballot, preference ballot, cumulated ballot,

Table 2.3. *Labels.*

<table>
<tr><td rowspan="2">Majoritarian methods</td><td rowspan="2"></td><td>absolute majority</td><td>AM</td></tr>
<tr><td>relative majority</td><td>RM</td></tr>
<tr><td rowspan="12">Proportional methods</td><td></td><td>single transferable vote</td><td>STV</td></tr>
<tr><td rowspan="8">Divisor</td><td>d'Hondt</td><td>DH</td></tr>
<tr><td>Equal Proportions</td><td>EP</td></tr>
<tr><td>Nohlen</td><td>NO</td></tr>
<tr><td>Belgian</td><td>BL</td></tr>
<tr><td>Danish</td><td>DN</td></tr>
<tr><td>Sainte-Laguë</td><td>SL</td></tr>
<tr><td>modified Sainte-Laguë</td><td>MSL</td></tr>
<tr style="display:none"></tr>
<tr><td rowspan="3">Largest Remainders</td><td>natural quota</td><td>NQ</td></tr>
<tr><td>Droop</td><td>DR</td></tr>
<tr><td>Imperiali</td><td>IMP</td></tr>
</table>

and ranking ballot, respectively; $\beta \in \{dis, reg, nat\}$, according to district, regional, or national level; β also contains the information whether the districts are single membered (*disS*) or multimembered (*disM*); $\gamma \in \{maj, pro\}$ (*maj* stands for majoritarian formula and *pro* stands for proportional).

At this point, we can catalogue all electoral systems, grouping them into classes distinguished by different strings. So, for example, the string "*list*|*disS*|*maj*" is the class of majoritarian electoral systems with single-membered districts where a list ballot is used (therefore, including the first-past-the-post and the double-ballot methods). In theory, any combination of the values of such variables generates an electoral system, but in practice only some sequences are feasible.

We can refine the above notation by adding more detailed features. To identify the precise majoritarian or proportional formula the electoral system uses, parameter γ is substituted by "$\gamma.x$" where the additional variable x is an abbreviation of the extended name. In Table 2.3 these abbreviations are listed.

The symbol "$\rightarrow$" separates any two consecutive electoral rounds. Within the same round, let the symbol "$\prec$" represent *priority* and the symbol "," represent *simultaneousness*. For example, "dis $\prec$ reg" means that in a first phase some seats are assigned at district level, and in a second phase the remaining seats are assigned at regional level. Instead "dis, reg" means that a fraction of the total number of seats is assigned at district level and another fraction at regional level.

There exists a correspondence between elements of a string on the same side of the priority and simultaneousness operators. For example, the string

$$dis \prec reg | pro.NQ \prec pro.DH$$

represents an electoral formula which assigns seats at district level with the natural quota proportional method and, then, it assigns the remaining seats at regional level with d'Hondt's proportional divisor method. Moreover, the notation "," can be used to represent mixed electoral systems, so that, for example, the string

$$\frac{1}{2} maj.RM, \frac{1}{2} pro.DH$$

denotes a mixed system in which one half of the total number of seats is assigned by plurality and the other half by d'Hondt's divisor method.

On the basis of these parameters we can already represent, with good approximation, nearly every electoral system in use, only with a string. Examples are provided in Table 2.4.

Table 2.4. *Examples of electoral systems in use (most information refers to Cox, 1997).*

Country	*Code*
Argentina	$\mathit{list} \mid \mathit{disM} \mid \mathit{pro.DH}$
Austria	$\mathit{list} \mid \mathit{disM} \prec \mathit{regM} \mid \mathit{pro.NQ} \prec \mathit{pro.DH}$
Belgium	$\mathit{rank} \mid \mathit{disM} \prec \mathit{reg.cartM} \mid \mathit{pro.NQ} \prec \mathit{pro.DH}$
Bolivia	$\mathit{list} \mid \mathit{disM} \mid \mathit{pro.NQ}$
Brazil	$\mathit{pref} \mid \mathit{disM} \mid \mathit{pro.NQ}$
Denmark	$\mathit{list} \mid \mathit{disM,nat.cartM} \mid \mathit{pro.MSL,pro.NQ}$
Finland	$\mathit{pref} \mid \mathit{disM} \mid \mathit{pro.DH}$
France	$\mathit{list} \mid \mathit{disS} \mid \mathit{maj.AM} \to \mathit{list} \mid \mathit{disS} \mid \mathit{mag.RM}$
Germany	$\mathit{list} \mid \mathit{disS,regM} \prec \mathit{nat.cartM} \mid \frac{1}{2}\mathit{mag.RM},\frac{1}{2}\mathit{pro.NQ}$
Greece	$\mathit{pref} \mid \mathit{disM} \prec \mathit{reg.cartM} \prec \mathit{nat.cartM} \mid \mathit{pro.DH} \prec \mathit{pro.NQ} \prec \mathit{pro.NQ}$
Iceland	$\mathit{list} \mid \mathit{disM} \prec \mathit{nat.cartM} \mid \mathit{pro.NQ} \prec \mathit{pro.DH}$
Ireland (Lower House)	$\mathit{rank} \mid \mathit{disM} \mid \mathit{pro.DR}$
Israel	$\mathit{list} \mid \mathit{natM} \mid \mathit{pro.DH}$
Italy	$\mathit{pref} \mid \mathit{dis.cartS,nat.cartM} \mid \frac{3}{4}\mathit{maj.RM},\frac{1}{4}\mathit{pro.NQ}$
Japan	$\mathit{list} \mid \mathit{disS,regM} \mid \frac{3}{5}\mathit{maj.RM},\frac{2}{5}\mathit{pro.DH}$
Luxembourg	$\mathit{pref} \mid \mathit{disM} \mid \mathit{pro.DH}$
The Netherlands (Second Chamber)	$\mathit{rank} \mid \mathit{natM} \mid \mathit{pro.DH}$
Norway	$\mathit{list} \mid \mathit{disM,natM} \mid \mathit{pro.MSL,pro.NQ}$
Portugal	$\mathit{list} \mid \mathit{disM} \mid \mathit{pro.DH}$
Spain (Congress of Deputies)	$\mathit{list} \mid \mathit{disM} \mid \mathit{pro.DH}$
Sweden	$\mathit{list} \mid \mathit{dis.cartM,nat.cart} \mid \mathit{pro.MSL,pro.MSL}$
Switzerland (National Council)	$\mathit{pref} \mid \mathit{disM} \mid \mathit{pro.DH}$
United Kingdom	$\mathit{list} \mid \mathit{disS} \mid \mathit{maj.RM}$
Uruguay	$\mathit{list} \mid \mathit{nat.cartM} \prec \mathit{dis.cartM} \mid \mathit{pro.DH}$
United States of America	$\mathit{list} \mid \mathit{disS} \mid \mathit{maj.RM}$

Chapter 3

The Performance of an Electoral System: Criteria and Indicators

Electoral systems can be evaluated and compared on the basis of their general performance and consequences in terms of technical and political features. We are interested in defining appropriate criteria and indicators to test the behavior of different electoral systems and to identify the variables that influence the consequences of the electoral outcome on the political structure of a country. We will examine the most popular indicators found in the literature and attempt to point out their advantages and disadvantages with the support of simple examples. These indicators can be classified according to the four phases of an electoral process, as described in Chapter 1. In particular, since the ridge of an electoral system is the transformation of votes into seats, we can make a distinction between pretransformation indicators (electoral participation, number of parties, electoral volatility), transformation ones (proportionality), and posttransformation ones (party power and coalitions, government stability).

The quantitative analysis presented here reflects a multidisciplinary perspective which takes into account the theory of political science as well as concepts of statistics and game theory.

3.1 Electoral Participation

By electoral participation a variety of phenomena can be intended. Especially in European countries, where the political tradition was inherited from the Greek $\pi o\lambda\iota\sigma$ and the Roman forum, the participation in political life starts far before the electoral campaign and the election itself. Nevertheless, opinion polls and newspaper articles are always concerned with the number of people who actually vote versus the number of those who stay at home. The most common indicator of electoral participation is, therefore, the share of votes cast:

$$\frac{\text{number of votes cast}}{\text{total number of voters}}$$

or, alternatively, the incidence of actual votes over the nonvoting electors, which can be calculated as follows:

$$\frac{\text{number of votes cast}}{(\text{total number of voters} - \text{number of votes cast})}.$$

In fact, several political scientists seem to use electoral participation as an indirect measure of the degree of democracy of a country. It has been observed that, in the Western world, electoral participation is higher when higher levels of education, social status, and income are taken into account. Nevertheless, some variables that depend on the structure of the political system and its institutions can also affect the degree of participation in the elections. In fact, abstention from voting can represent a marginal occurrence or it can be a significant phenomenon in different countries although they have the same social and economic background. In such context, Jackman (1987) suggests a list of elements, strictly connected to the electoral system adopted, which can be considered important for electoral participation:

- the degree of competition in each district;
- the degree of proportionality of the vote-seat transformation;
- the number of parties;
- the number of governing Chambers;
- the practice of compulsory vote.

In fact, when the parties must compete in large districts where a proportional seat assignment method is adopted, they must also battle in electoral campaigns spread on the whole national territory and encourage their supporters to vote in every district. On the other hand, when single-member districts are considered, electoral campaigns will be concentrated only in the districts where the winner is not already determined by a sufficiently high concentration of electors belonging to the same party. The other electors, who are aware of the fact they will lose, might avoid casting a "useless" vote. The same kind of discouragement can be found when there is only a small number of parties or alternatives from which to choose. This is usually associated with majoritarian methods, rather than with proportional ones, since with the latter even small and scattered parties have a greater chance to win a seat. Therefore, the degree of proportionality of the electoral formula adopted can enhance or reduce the number of votes cast by guaranteeing that each vote will actually be translated into some (although small) amount of representation in the Parliament.

On the other hand, when the number of parties is too large, political agreements and coalitions will be necessary to produce a government. In this case voters may be disoriented by the fact they do not directly determine who will actually hold the decisive power, so they may not feel effective. Higher levels of electoral participation have been registered in countries where there is a unique Chamber. The reason might be that a unique Chamber concentrates the executive power making any elector aware of the decision-making process and those responsible. In some countries vote is compulsory, and in this case there is no point in measuring electoral participation. However, when the penalties for not voting are weak, abstention can still exist.

Many other factors could be taken into consideration; however, it is already clear that the kind of electoral system adopted cannot be ignored in the analysis of electoral participation. In particular in Europe, proportional systems are usually considered to encourage a vast participation. But, in fact, the features that can be used to explain the inclination to abstention depend on the electoral system examined. As we have already stated, when proportional systems are considered, since the seats assigned in a district visibly reflect the votes cast, electors will feel the vote can really have an effect on the final result, even when their preference goes to a very small party. In addition, political parties under proportional systems tend to prepare widespread electoral campaigns since all districts are competitive and there is always a chance to win a seat. Finally, these systems are usually associated with a large number of political parties, which allows a larger number of citizens to find a party representing their political ideal.

On the other hand, when a plurality system is adopted, the fact that the vote-to-seat transformation process is very easy to understand might encourage participation because of its simplicity. Moreover, these systems are usually associated with a small number of parties, increasing the possibility of single-party governments. In this case the electors will recognize the fact that the government itself is directly determined by their vote rather than by the political agreement made between the parties after the vote.

Double ballot lies somewhat between the two systems mentioned above. With respect to plurality it is less disadvantageous for the smaller parties, and with respect to proportional systems it requires that coalitions between parties be made before and not after the vote. So, depending on the electoral system, different features might encourage vote participation, but the political tradition, the evolution of society, the structure of the governing institutions, and the history of a country cannot be put aside in the analysis of this kind of phenomena.

3.2 The Number of Political Parties

The number of parties that constitute the political system has always been considered an important parameter to describe the complex network of competitive relationships between the different actors of the system and, in particular, between the parties, the institutions, and the electors themselves. Although it is a basic variable in the description of an electoral system, just counting the

Table 3.1. *Different vote distributions in a three-party system.*

Parties	*System* 1	*System* 2	*System* 3
A	90%	45%	34%
B	9%	40%	33%
C	1%	15%	33%

number of parties that go to the elections does not seem to be the best way to define how many competitive positions are to be considered in a given political system.

Many different indicators have been proposed to count the number of parties in a system, and they all belong to the pretransformation phase, since, even if they can be strongly linked to the electoral formula adopted, they typically concern the vote distribution before seats are assigned.

The number n of parties that are competing in a given election is very simple to determine, but it tells nothing about the relative strength of these parties and their consensus among the voters. The simple number of parties fails to distinguish between radically different party systems such as those described in Table 3.1.

In the example the number of parties does not change from one system to another, but one does not need to be an expert to see that the three systems represent totally different competitive relationships. System 1 can in fact be considered a single-party system since party A enjoys nearly all of the vote; system 2 shows a case in which two parties (A and B) are more or less on even terms and a smaller but still consistent party exists (C); in system 3 all parties show the same relative strength.

Henceforth we shall employ the following notation: $\omega_i = \frac{v_i}{\mathcal{P}}$ is the vote share of party $i = 1, 2, \ldots, n$, and $\sigma_i = \frac{s_i}{S}$ is the seat share of party $i = 1, 2, \ldots, n$. If $\boldsymbol{x} = (x_1, x_2, \ldots, x_n)$ is an n-vector, then $(x_{[1]}, x_{[2]}, \ldots, x_{[n]})$ denotes the n-vector with the same components as x but sorted in nonincreasing order: $x_{[1]} \geq x_{[2]} \geq \cdots \geq x_{[n]}$.

A way to emphasize the relative strength rather than the simple number of parties could be to consider the vote share of the strongest party, $\omega_{[1]} = \frac{v_{[1]}}{\mathcal{P}}$ or the combined vote share of the two most-preferred parties, $\omega_{[1]} + \omega_{[2]} = \frac{v_{[1]}}{\mathcal{P}} + \frac{v_{[2]}}{\mathcal{P}}$. This tells us how closely a system approaches the condition of a one-party or a two-party system. But, unless they are reported together, these indicators say nothing about how evenly the stronger parties compete with each other: they only measure the degree to which they can, together, dominate all the others.

For this reason it seems sensible to introduce an indicator capable of describing the general structure of the competitive relationships, including all of the parties that participate in the election. This indicator is usually referred to as fractionalization (as opposed to concentration) of vote shares. A nonfractionalized system is a one-party system where, in fact, there is no place for any kind of competition, while a highly fractionalized system is made up of several parties that all compete with the same strength (or share of votes).

The idea underlying fractionalization is to count the number of party shares and to evaluate their relative equality. The traditional method used to measure fractionalization is based on computing the probability that two randomly selected voters do not vote for the same party [Rae, 1967]. If a unique party holds 100% of the total vote, the probability that two voters do not agree on the party to vote for is zero. As the number of parties in the system increases, the fractionalization does, too, but it is equal to its maximum theoretical value (i.e., one) only if there are as many parties as the electors.

The chance that any two voters will have chosen the same party is given by

$$\sum_{i=1}^{n} \omega_i^2,$$

so that vote fractionalization (or the probability of dyadic disagreement) is its complement:

$$F = 1 - \sum_{i=1}^{n} \omega_i^2.$$

The indicator F takes values in the $[0,1]$ interval, and it is sensitive to both the number of parties and the relative equality of the party shares.

The idea of fractionalization has been exploited by Laakso and Taagepera (1979) to count what they call the effective number of parties. This indicator represents the number of parties of equal strength (i.e., with the same vote share) in a theoretical political system that is equivalent to the one under consideration, that is, with the same degree of vote fractionalization (according to Rae's index F). In fact, a real political system does not necessarily consist of parties having relatively equal strengths.

Let N_α be the general indicator for the number of parties in a political system, where vote shares are considered

$$N_\alpha = \left(\sum_{i=1}^{n} \omega_i^\alpha\right)^{\frac{1}{1-\alpha}}.$$

Now, when $\alpha = 2$ this is the Laakso–Taagepera index; when α approaches 1, this indicator approximates another index called the Wildgen index (1971) which is based on the concept of entropy and is precisely equal to

$$N_1 = \exp\left(-\sum_{i=1}^{n} \omega_i \log \omega_i\right).$$

Wildgen's index is also known as the hyperfractionalization index since it is even more sensitive than the others to the number of parties n. In fact, when the parameter α tends to 0, the general indicator N_α tends to n, while it tends to overemphasize the weight of the largest vote share $\omega_{[1]}$ when α tends to infinity.

On the other hand, the inverse of the number of parties $\frac{1}{n}$ represents the level of political concentration of the system and its complement to one, $\left(1-\frac{1}{n}\right)$, can be used as another and simpler fractionalization index. Following the same generalization process as above, a family of indexes for fractionalization can be obtained as follows:

$$F_a = 1 - \frac{1}{N_\alpha}$$

for different values of α.

Yet, recently, Molinar (1991) proposed a new indicator to measure the effective number of parties that does not belong to the family of indexes discussed above, although it uses Laakso and Taagepera's index:

$$NP = 1 + \frac{\left(\sum_{i=1}^{n} \omega_i^2\right) - \omega_{[1]}^2}{\left(\sum_{i=1}^{n} \omega_i^2\right)^2} = 1 + N_2 - N_2^2 \omega_{[1]},$$

where $\omega_{[1]}$ is the largest vote share. Like Rae's index, NP has an interpretation in terms of probability. It compares the probability that two randomly selected voters belong to a minority party (that is, any party that does not coincide with the strongest in terms of vote shares) with the probability that they belong to the same party, including the strongest.

The best way to grasp the different performance of the indexes that have been introduced is to consider a few examples. Table 3.2 describes five political systems, with their corresponding number of parties and vote shares.

Table 3.2. *Different party scenarios.*

System	*Party* 1	*Party* 2	*Party* 3	*Party* 4
a	1.00	-	-	-
b	0.80	0.20	-	-
c	0.50	0.50	-	-
d	0.33	0.33	0.34	-
e	0.25	0.25	0.25	0.25

In Table 3.3 Rae's fractionalization F and Wildgen's index N_1 have been computed for each one of the systems described above.

Notice that F is sensitive to the number of parties and their relative strength since systems b and c yield a different value, although they both represent two-party systems. On the other hand, a proportional increase in the degree of

Table 3.3. *Rae's index versus Wildgen's index.*

System	F	N_1
a	0	1
b	0.32	1.64
c	0.50	2
d	0.67	3
e	0.75	4

competition is not reflected in a proportional increase of the value of this index. In fact, by comparing systems c and e where all parties have equal vote shares and thus the same strength, although the vote dispersion increases twice as much, F increases about 50%, from 0.5 to 0.75.

Wildgen's hyperfractionalization index is sensible both to the number of parties and to the vote distribution, but it increases with respect to the latter as a linear function. In this case, the index goes from 2 to 4, i.e., it increases proportionally, when systems c and e are considered. The maximum value that N_1 can assume is n.

To compare this index with the Laakso and Taagepera index N_2 and Molinar's index NP, consider the following example (Table 3.4):

Table 3.4. *Comparison between the Wildgen, Laakso–Taagepera, and Molinar indexes.*

System	***Party***							N_1	N_2	NP
	1	2	3	4	5	6	7			
a	1.00	-	-	-	-	-	-	1	1	1
b	0.50	0.50	-	-	-	-	-	2	2	2
c	0.33	0.33	0.34	-	-	-	-	3	3	3
d	0.51	0.42	0.05	0.01	0.01	-	-	2.58	2.28	1.93
e	0.51	0.26	0.11	0.11	0.01	-	-	3.41	2.84	1.74
f	0.40	0.37	0.11	0.11	0.01	-	-	3.55	3.11	2.56
g	0.40	0.37	0.09	0.09	0.05	-	-	3.73	3.17	2.56
h	0.51	0.49	-	-	-	-	-	2	2	1.96
i	0.70	0.05	0.05	0.05	0.05	0.05	0.05	3.15	1.99	1.06

When all parties equally divide the total vote, the three indexes simply yield the number of parties in competition. These values are 1, 2, and 3 for systems a, b, and c, respectively.

Now consider systems f and g, which both consist of two strong parties and three minor parties with different vote shares. Both N_1 and N_2 produce values above 3, but the two political systems are basically two-party systems, as is detected by NP.

On the other hand, we could be surprised by noticing the difference between N_1 and N_2 when systems h and i are examined. The vote distribution in these two political systems is very different, since in the first case two parties are almost equally divided, while in the second case the first party is highly concentrated and numerous other small parties have the same (insignificant) relative strength. In other words, h is an example of a two-party system, while i is an example of a one-party system: all three indicators correctly represent system h. On the other hand, N_2 is closer to interpret case i than N_1, but only NP seems to fully capture it.

Generally, what happens is that N_1 tends to be too sensitive to the total number of parties, even when they are very small, while N_2 is generally too sensitive to the fraction of votes cast for the strongest party. This can be pointed out by closely examining system f, as in Table 3.5, where the contributions due to the single parties are detailed.

Table 3.5. *The contribution of single parties to N_1 and N_2 in system f.*

Party	ω_i	ω_i^2	$-\omega_i \log \omega_i$
1	0.70	0.4900	0.25
2	0.05	0.0025	0.15
3	0.5	0.0025	0.15
4	0.05	0.0025	0.15
5	0.05	0.0025	0.15
6	0.05	0.0025	0.15
7	0.05	0.0025	0.15
total	1	0.5050	1.15

The third column of this table is the square of the vote share of each party (needed in computing N_2). The strongest party's contribution to the index represents about 97% of the total, so that the first party weighs $0.97 \times 1.99 = 1.93$ in the Laakso–Taagepera index. A similar analysis can be made to see that the first party's contribution to Wildgen's index is only 22%, corresponding to a total weight of about 0.68, while the smaller parties all weigh about 0.41 (each representing 13% of the total in this case).

This example was used by Molinar to introduce his index NP which, in many of the cases examined above, does seem to perform better with respect to our intuition. This advantage is mainly due to the fact that NP isolates the strongest party, assigning it a weight equal to 1, regardless of its relative strength.

Nevertheless, in some situations the superiority of NP might be controversial. In fact, suppose three parties constituting a political system evolve in time in terms of vote shares, so that the third party progressively gains votes while the first party remains with the same share (Table 3.6).

Table 3.6. *Values of the Laakso–Taagepera index and the Molinar index when a third party is introduced in the system.*

Period	*Party* 1	*Party* 2	*Party* 3	N_2	NP
t	0.55	0.45	-	1.98	1.79
$t+1$	0.55	0.35	0.10	2.30	1.70
$t+2$	0.55	0.25	0.20	2.47	1.62

While N_2 detects the increase in the number of parties and in the degree of competition by producing increasing values in the three time periods, NP decreases. On the other hand, since the strongest party has a strong opponent only at time t and clearly dominates all the others at time $t+2$, the interpretation underlying the NP index, i.e., that of a one-party system, might be preferred.

3.3 Electoral Volatility

After introducing indicators that measure the level of political participation and the number of alternatives or parties the electors can vote for, another class of pretransformation indicators, concerning the change of the choice made by the electoral body in time, must be considered.

Electoral volatility is a dynamic concept that tries to explain how much the competitive outcome of the present election reflects the situation expressed in earlier elections and, in particular, if there is a radical alteration with respect to the past or if the party positions are more or less unchanged.

Bartolini and Mair (1992) identify a set of variables that affect the electoral volatility registered in time and in space. Most of them are social variables, including electoral participation, alone or together with the degree of social difference encountered in the population. Others are institutional variables, including the number of political parties, their degree of ideological or political diversity, and the method used to translate votes into seats.

The importance of appropriately measuring electoral volatility is due to the fact that it is not only affected by some features of a political system, but it can itself determine significant changes in the evolution of this system. High volatility between successive elections can have surprising results on the final government, since it will require new coalitions and political agreements to be made. This will necessarily occur when the volatility is high enough to radically change the seat distribution.

The connection between electoral volatility and the features mentioned above are illustrated in Figure 3.1. The left-hand side shows the factors affecting electoral volatility, while the right-hand side shows the factors on which it acts.

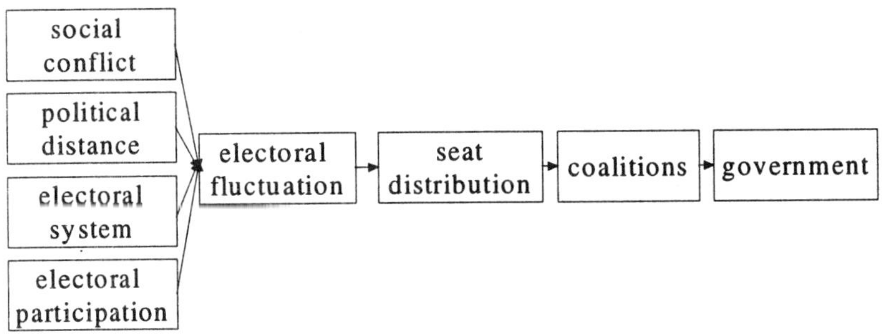

Fig. 3.1. *Interrelationship between electoral volatility and the political system.*

The simplest measure for electoral volatility is the average change of votes between two consecutive elections, at time t and $t+1$:

$$EF = \sum_{i=1}^{n} \frac{|v_i(t) - v_i(t+1)|}{2},$$

where $v_i(t)$ represents the total amount of votes cast for party i at time t. This index is precisely the sum of the absolute differences between the votes each party has received in the two elections, divided by two since everything that is gained by one party is always a loss for the others. For example, consider three parties A, B, and C who received 30, 40, and 30 votes, respectively, in the past elections, while they receive 28, 43, and 29 votes, respectively, in the present. The electoral body is slowly moving toward the central party B as is registered by the indicator above, which, in this case, is equal to 3. However, this index does not actually show the direction of the changes that have occurred in the vote distribution but only the number of votes that have changed.

In fact, in some cases it might be useful to know the degree of electoral volatility, which is the main cause of the change in the vote distribution. Consider the same example, but suppose that at time $t+1$ the vote distribution has not changed. Then the volatility indicator will be equal to zero, but this does not automatically imply that the electors have all cast their vote for the same party they voted for at time t. Even if all of the electors have changed their vote, the final balance could be the same.

Let us now consider a *vote transition matrix*. The entries of this matrix, t_{ij}, represent the percentage of votes that were assigned to party i at time t and have moved to party j at time $t+1$. The size of the matrix is $m \times n$, where m is the total number of parties competing at time t and n the total number of parties competing at time $t+1$. In the example, suppose it is possible to count the quotas of electors that have migrated from one party to another and let the result be

$$T = \begin{bmatrix} 0.30 & [9] & 0.50 & [15] & 0.20 & [6] \\ 0.15 & [6] & 0.40 & [16] & 0.45 & [18] \\ 0.50 & [15] & 0.30 & [9] & 0.20 & [6] \end{bmatrix},$$

where the numbers of migrating voters are in brackets. Notice that, even though the total amount of votes A, B, and C receive is still equal to 30, 40, and 30, respectively, the people voting for these parties have changed significantly in time. Only 20% of C's supporters remain with this party at time $t+1$, while 50% of them have preferred A this time. If the electoral system does not require the use of blocked lists (i.e., the voter gives a vote to the most preferred party and a preference to a specific candidate), the new political scene will usually be characterized by the same parties but by different candidates, causing a shuffle and a reorganization to be made within the parties themselves.

However, the indicator introduced above is not sensitive to the direction of electoral volatility, so we suggest using a specific index that also takes into consideration the fact that it should be "easier" for a voter to change his or her vote in favor of a party under the condition that the new party is politically close to the initial one. In other words, we are interested in evaluating to what extent the changes are radical in the sense that they concern swapping voters between parties that are politically very far from each other. To do this, we suppose that the parties can be ordered from left to right. Then a symmetric matrix Δ can be computed to represent the distance between pairs of parties, where each element d_{ij} is equal to the number of steps separating party i from party j in the ordered list (thus, $d_{ii} = 0$ for all parties i, $d_{ij} = 1$ if party j is next to party i, and so on). To measure the degree of vote transformation we can therefore compute the following index:

$$E = \frac{1}{\bar{d}} \sum_{i=1}^{m} \sum_{j>i}^{n} \left| t_{ij} v_i(t) - t_{ji} v_j(t) \right| d_{ij},$$

where t_{ij} represents the quota of voters who have changed their vote from the former to the latter party at time $t+1$ and $\bar{d}$ is the maximum row total in the Δ matrix, $\bar{d} = \max_i\{\sum_{j>i} d_{ij}\}$. Notice that if party i does not compete at time t or if party j does not compete at time $t+1$, we consider t_{ij} to be equal to 0.

In the case of the example, supposing that A, B, and C are already ordered from left to right according to their political position, the distance matrix Δ will be equal to

$$\Delta = \begin{bmatrix} 0 & 1 & 2 \\ 1 & 0 & 1 \\ 2 & 1 & 0 \end{bmatrix},$$

and $E = 12$. However, more specific examples are useful to better understand the way such an indicator works.

First, let us suppose a party (called D) has disappeared from the political scene between time t and $t+1$. The transition matrix is as follows:

$$T = \begin{bmatrix} 0.8\,[16] & 0.1\;\,[2] & 0.1\;\,[2] & 0\;[0] \\ 0.1\;\,[3] & 0.7\,[21] & 0.2\;\,[6] & 0\;[0] \\ 0\quad\;[0] & 0.3\,[12] & 0.7\,[28] & 0\;[0] \\ 0.2\;\,[2] & 0.2\;\,[2] & 0.6\;\,[6] & 0\;[0] \end{bmatrix},$$

and the parties are ordered from left to right, from A to D. The total vote cast for each party has necessarily changed between time t (where it was equal to 20, 30, 40, and 10 for A, B, C, and D, respectively) and $t+1$ (where it is equal to 21, 37, 42, and 0) since one of the parties has disappeared. Notice that $EF = 10$ while $E = 4.5$.

Now let us suppose that a new party appears between the two elections and that such party, called X, can be considered to be between B and C in the political rank list. The transition matrix that has been registered is equal to

$$T = \begin{bmatrix} 0.6\,[18] & 0.3\;\,[9] & 0.1\;\,[3] & 0\quad\;[0] \\ 0.2\;\,[6] & 0.7\,[21] & 0.1\;\,[3] & 0\quad\;[0] \\ 0\quad\;[0] & 0\quad\;[0] & 0\quad\;[0] & 0\quad\;[0] \\ 0\;\,[20] & 0.1\;\,[4] & 0.2\;\,[8] & 0.7\,[28] \end{bmatrix}.$$

The electoral volatility has significantly increased with respect to the last example ($EF = 18$, here since A, B, X, and C, respectively, gained 30, 30, 0, and 40 votes at time t and 24, 34, 14, and 28 votes at time $t+1$), but the degree of vote change is nearly the same as the one registered before (since $E = 4.7$).

3.4 Measuring Proportionality

Indicators that try to evaluate the degree of proportionality of an electoral system are based on the comparison between the quota of votes and the quota of seats each party obtains. The electoral formula in itself is not enough to explain the difference between vote and seat quotas. Other variables, and in particular district magnitude, can create distortion as well. In fact, different degrees of proportionality can be registered in the same election given a specific electoral formula by modifying the size of the electoral districts which subdivide the national territory.

Of course, whatever the combination of formula and district size, some distortion between quota of votes and seats will always exist, mainly because the number of seats that can be assigned to each party must be an integer. In fact, if 5 seats were to be distributed among 3 parties with 60, 28, and 12 votes, exact proportionality would require that 3 seats be assigned to the first party, 1.4 to the second, and 0.6 to the last—and this is obviously impossible.

In many electoral systems, a further obstacle to the achievement of proportionality is due to the introduction of exclusion thresholds, i.e., vote quotas below which a party does not qualify for the seat distribution. In such case disproportionality is usually a necessary consequence of the purpose of decreasing the number of parties on the political scene. The interest accorded to the analysis of proportionality is reflected in the effort dedicated to the construction of a suitable measure of disproportionality. Such task is much more problematic than it may seem, as will be explained in Chapter 6.

Let ω_i and σ_i be the share of votes obtained and the share of seats assigned to party i, respectively, where $i = 1, 2, \ldots, n$. One of the first and still most cited indicators is due to Loosemore and Hanby (1971), which is computed as the sum of the absolute differences between vote and seat shares, divided by 2:

$$LH = \frac{\sum_{i=1}^{n} |\omega_i - \sigma_i|}{2}.$$

This is equal to 0 if and only if $\omega_i = \sigma_i$ for every party $i = 1, 2, \ldots, n$, therefore, when there is exact proportionality. One of the disadvantages of such an index is its sensitivity to the number of parties. In fact, in front of a very splintered political scene where a large number of relatively small parties coexist, the value of the Loosemore–Hanby index will necessarily increase with the number of parties even if some of them are so small that they should be considered insignificant.

Another disproportionality index is due to Rae (1967). It attempts to cut out parties that can be considered too small to effectively count by introducing a threshold of 0.5% of the total votes and, in the measurement of proportionality, only those parties that are above the threshold are considered. The sum of the absolute differences between vote and seat shares of such parties is then divided by their number:

$$R = \frac{\sum_{i \in \bar{I}} |\omega_i - \sigma_i|}{|\bar{I}|},$$

where $\bar{I} = \{i : \omega_i > 0.005\}$ is the set of parties that is above the fixed threshold. Again the index is equal to 0 in the case of exact proportionality. The Rae index distinguishes between cases where the distortion of the vote-seat shares is due to many parties each producing small differences with respect to those where it is due to a small number of parties all yielding large differences. This distinction is not always possible when the Loosemore–Hanby index is used.

However, Rae's index can be criticized because in a system where there are many parties that are just under 0.5% of the vote total, it will tend to assume very low values, making the system look more proportional than it actually is.

To solve this problem, Lijphart (1990) proposed dividing the sum of the absolute differences between vote and seat shares by the effective number of parties or to use the average disproportionality between the two largest parties to represent the disproportionality of the whole system:

$$L = \frac{|\omega_{[1]} - \sigma_{[1]}| + |\omega_{[2]} - \sigma_{[2]}|}{2}.$$

The indexes listed above all measure disproportionality; that is, they are equal to zero when the system is perfectly proportional.

There are many other measures that are intended to be approximately halfway between LH and R. The most common is called the least square index, and it is defined as follows:

$$S = \sqrt{\frac{\sum_{i=1}^{n}(\omega_i - \sigma_i)^2}{2}},$$

which, as the other, takes values in the $[0, 1]$ interval. The interesting feature of this index is that it tends to assign a larger weight to vote-seat differences that are big rather than to those that are small. When there are only two parties that compete in the system, S coincides with L and R.

The indexes considered up to now are all based on measuring the difference between vote and seat shares. Another class of indexes based on the ratio between vote and seat shares can be considered, too. When absolute differences are adopted, the distortion due to assigning 55% of the seats to a party that has obtained 50.1% of the votes is the same as that due to assigning 5% of the seats to a party with only 0.1% of the votes. But, intuitively to most people, in the first case the party is only slightly overrepresented, while in the second it is grossly overrepresented. This problem can be solved by referring to relative differences (or ratios) instead, so that the overrepresentation in the first case is equal to 1.1 and in the second is equal to 50.

This class includes indexes known as the Sainte-Laguë, the d'Hondt, and the equal proportions index. Let v_i and s_i be the number of votes obtained and the number of seats assigned to party i, respectively, and let P and S be the total number of votes and seats. Then the Sainte-Laguë index measures the squared differences between the $\frac{s_i}{v_i}$ ratio and the total seat-to-vote ratio, where each difference is weighted by the number of votes the corresponding party has obtained:

$$SL = \sum_{i=1}^{n} v_i \left(\frac{s_i}{v_i} - \frac{S}{P}\right)^2.$$

This index is equal to 0 (exact proportionality) when $\frac{s_i}{v_i} = \frac{S}{P}$ for every party $i = 1, 2, \ldots, n$. One shortcoming of the Sainte-Laguë index is that it is not defined when at least one of the parties has not obtained any vote (this is obviously a purely theoretical argument, but some electoral formulas, such as Adams's, would do this). In any case, the Sainte-Laguë index has no upper bound, making its interpretation more difficult.

The equal proportions index takes into account seat and vote shares as follows:

$$EP = \sqrt{\sum_{i=1}^{n} \frac{\omega_i^2}{\sigma_i} - 1}.$$

Unfortunately it is quite useless for practical purposes, since if a party obtains no seats it will give an infinitely large contribution to the total sum.

Finally, the d'Hondt index simply measures the seat-to-vote ratio of the most overrepresented party; that is,

$$H = \max_{i=1,2,\ldots,n} \frac{\sigma_i}{\omega_i}.$$

Its minimum value (exact proportionality) is 1, when all parties have identical seat-to-vote ratios, and its maximum value, attained if a party with no votes receives some seats, is (conventionally) plus infinity. But the real disadvantage of such an index is that it is too sensitive to the overrepresentation produced by small parties. In some cases it would be preferable to consider only the seat-to-vote ratios of those parties that have received at least 0.5% of the total vote.

3.5 Party Power and Coalitions

The most significant contribution to the analysis of the behavior of political coalitions is due to decision and game theory. The models commonly used in this field are usually based on the theory of rational actors. In fact, every player of the game will act according to a specific utility function in order to maximize his gain. This assumption can be extended to the way each party will act once it has won the election and it must decide whether to join other parties in a coalition.

Different theories in the analysis of coalitions generally disagree on the criterion that should define a coalition, that is, the principles that rule the merging of parties. A possible approach focuses on the political or ideological features involved in making a coalition and its consequences. The approach we will adopt is merely quantitative and relies only on numerical data by examining the technical possibility and convenience to form a coalition but ignoring the related political attributes.

In fact, the only kind of coalitions we are interested in are *winning coalitions*, that is, those that represent more than half of the seats in the Parliament. Let C be a set of parties selected in $I = \{1, 2, \ldots, n\}$ and $s(C) = \sum_{i \in C} s_i$ the corresponding number of seats; then the set of winning coalitions W is given by

$$W = \left\{ C : s(C) > \frac{S}{2} \right\}.$$

In most cases it is not sufficient for a party to know that the coalition is a winning coalition to actually join it. In fact, there are other criteria which are taken into account while evaluating the convenience for a party to join a coalition. In the literature there are four main criteria that govern the aggregation of parties, and each one defines a different subset of W.

The set of minimal winning coalitions contains all winning coalitions that would not be so if any party decides to drop out:

$$C_m = \left\{ C : C \in W, \;\; s(C - \{i\}) < \frac{S}{2} \;\; \forall i \in C \right\}.$$

On the other hand, we may consider winning coalitions with the smallest number of seats. These usually form a subset of C_m and are called the most economic winning coalitions [Riker, 1962], defined as follows:

$$C_s = \{C : C \in W,\ s(C) \leq s(C')\ \ \forall C' \in W\}.$$

Instead of considering the number of seats, we may be interested in the number of parties composing the coalition and, therefore, in the set of winning coalitions that contain the smallest number of parties. This criterion is usually known as the bargaining proposition [Leiserson, 1968] and defines the following subset of W:

$$C_e = \{C : C \in W,\ |C| \leq |C'|\ \ \forall C' \in W\}.$$

If the political nature of the parties makes some coalitions incompatible, we can restrict our attention to a set of "politically" feasible coalitions. In this case the political or ideological distance between pairs of parties must be measured to distinguish between their possible combinations [Leiserson, 1966].

Various distances have been proposed, and they are all based on the order of the different parties along the left to right political axis. In fact, suppose that five parties can be ordered from left to right, according to their political proposals, as follows: A, B, C, D, E. Then each party can be assigned a rank from left to right; that is, A= 1, B= 2, etc. The *Leiserson distance* between two parties is the sum of the number of parties between them and the number of positions the first party must move to reach the second (the number of "steps" between the two). For example, the distance between A and E, d(A,E), is equal to 7 because there are three parties between them (B, C, and D) and A would have to move four steps to reach E. Similarly, d(A,B) is equal to one. Following this kind of argument, the distance of a coalition is defined as the sum of the distances between the successive parties and is composed by, for example, d(ACD)= d(A,C)+d(C,D)= 4.

If $r(i)$ and $r(j)$ represent the rank of party i and j, respectively, and, say, $r(j) > r(i)$, then the distance between the two parties can be calculated easily by using the following expression:

$$d(i,j) = 2(r(j) - r(i)) - 1.$$

The set of minimum distance coalitions is therefore given by

$$C_d = \{C : C \in W,\ d(C) \leq d(C')\ \ \forall C' \in W\},$$

where $d(C)$ is the distance of the coalition as defined above.

The different subsets of W that have been considered do not necessarily coincide. Consider five parties that go from A to E in left-to-right order and control 22, 15, 18, 25, and 20 seats, respectively (for a total of 100 seats). A winning coalition is a set of parties that can add up to more than 50 seats. In particular,

$$W = \left\{ \begin{array}{l} \text{ABCDE(100)}, \\ \text{ABCD(80), ABCE(75), ACDE(85)}, \\ \text{BCDE(88), ABC(55), ABD(66), ABE(57), ACD(65)}, \\ \text{ACE(60), ADE(67), BCD(58), BCE(53)}, \\ \text{BDE(60), CDE(63)} \end{array} \right\},$$

while

$$\begin{aligned} C_m &= \left\{ \begin{array}{l} \text{ABC(55), ABD(66), ABE(57), ACD(65)}, \\ \text{ACE(60), ADE(67), BCD(58), BCE(53)}, \\ \text{BDE(60), CDE(63)}, \end{array} \right\}, \\ C_s &= \{\text{BCE(53)}\}, \\ C_e &= C_m, \end{aligned}$$

where the number of seats controlled by each coalition is given in brackets. After calculating the distance corresponding to each winning coalition, we can also observe that

$$C_d = \{\text{ABC(55), CDE(63)}\}.$$

Since it may be impossible to find a coalition satisfying all of these criteria simultaneously, we are interested in identifying coalitions that are nondominated. For example, if the following parties are considered in their left-to-right order (the corresponding number of seats is in brackets)

A(24), B(40), C(36),

there are four winning coalitions, namely, AB with 64 seats and distance equal to 1, AC with 60 seats and distance 1, BC with 76 seats and distance 1, and ABC with 100 seats and distance 2. Now suppose that party A and party B both evaluate these coalitions on the basis of the most economic winning coalition C_s first, while party C considers the minimum distance criteria C_d first. Then party A will adopt the following preference order $\prec$ on the winning coalitions:

$$\text{AC} \prec \text{AB} \prec \text{ABC} \prec \text{BC},$$

while party B will prefer

$$\text{AB} \prec \text{BC} \prec \text{ABC} \prec \text{AC},$$

and finally party C will consider

$$\text{BC} \prec \text{ABC} \prec \text{AC} \prec \text{AB}.$$

In this case it is impossible to identify a nondominated coalition. In fact, the AB coalition is dominated because both party A and party C would prefer the coalition AC; but AC is dominated, too, since both party B and C would prefer BC; moreover, BC is dominated since both A and B prefer the combination AB; and, finally, ABC is dominated because A and B prefer AB and B and C prefer BC.

In this simple example, each party has selected a criterion to define its preference for the set of winning coalitions. When more criteria are considered simultaneously, it can be shown [Taylor, 1973] that nondominated coalition can be found only if each party lists the winning coalitions in the same lexicographic order. When two alternatives are ordered in the lexicographic order, they are first evaluated on the basis of the first criterion, say, c_1, and if they coincide in terms of c_1, they are then judged on the basis of c_2, and so on. The order in which the different criteria are taken into account is essential, since different sequences can produce radically different results.

The analysis carried out up to now is a static analysis of coalition formation since it only concerns the result of each coalition in terms of the number of seats it can control without considering the dynamic aspects of this phenomenon, mainly due to the fact that some parties might take an initiative before the others. A dynamic approach is attempted by the model known in literature as the share of spoils model [Brams, 1975].

The *share of spoils* of a party is the probability that such party makes a coalition a "winning" one by joining it. In this kind of model, nonwinning coalitions compete to try to convince other parties to join them so that the total number of seats exceeds the 50% threshold.

Consider four parties, A, B, C, and D, which hold 42, 32, 16, and 10 seats, respectively. If A(42), B(32) are considered as competing nonwinning coalitions and C and D as possible allies, then the alternatives in terms of winning coalitions are AD(52), AC(58), ADC(68), and BCD(58). All coalitions that simultaneously involve A and B must be excluded from the analysis since these two parties are considered as competing coalitions. If no other information is provided, we must assume that these four winning coalitions occur with the same probability, so that the probability that a coalition built on A will win is 0.75 and the probability a coalition built on B will win is 0.25.

Now examine the alternatives party D can choose from. If D decides to join A, the probability that A wins with D is 1 since AD is a winning coalition, but D's contribution is only 0.25 (that is $1 - 0.75$). On the other hand, if D decides to join B, the possibility that a winning coalition comes out depends on C's choice. In particular, the probability that the coalition with B will be a winning coalition is 0.25, and D's contribution is 0.5 since the probability that C joins BD and the probability it joins A is the same.

The share of spoils of the possible coalitions arising from D's choice (that is AD and BD) represents the contribution of party D to the new coalition weighted by the probability that such coalition is a winning coalition. In this case, we have the share of spoils $ss(\text{AD}) = 0.25 \times 1 = 0.25$ and $ss(\text{BD}) = 0.25 \times 0.5 = 0.125$. Since D is a rational actor, it will try to maximize the probability that the coalition it joins is a winning coalition, and therefore it will choose to join A. In some ways, the idea underlying the share of spoils is that each party will consider not only those coalitions that make it part of the winners but also how important its contribution is in making the coalition a winning one.

Another topic of great interest is the relationship between different parties that belong to the same coalition. In fact, some parties may turn out to be more influential than others and not necessarily because they enter the coalition with a larger number of seats. The analysis of the power distribution among the parties of a coalition is essentially based on the concept of voting power, for which numerous indicators have been proposed in the literature.

Generally, the members that can actually influence or control the final outcome of a group are the *real actors*, while the others are called *dummies*. In the case of coalitions, the final outcome can be viewed as the political action to be undertaken in Parliament. In some cases, there could be only one real actor in the coalition, which will therefore be called a *dictator*.

The power or weight of the different actors in a coalition is one of the factors used to evaluate the convenience perceived in joining it. A useful notion in order to evaluate voting power is the characteristic function of a game.

In our context, the *characteristic function* is a real-valued function χ defined on the set of all coalitions such that

$$\begin{aligned} &\chi(\emptyset) = 0, \\ &\chi\left(C \bigcup C'\right) \geq \chi(C) + \chi(C') \end{aligned}$$

for all pairs of coalitions C, C'. In other words, the characteristic value of an empty coalition is zero and the function is superadditive, so that if any two coalitions decide to join into one, their value certainly does not decrease.

The simplest case corresponds to the following characteristic function, which only distinguishes between winning and losing coalitions:

$$\chi(C) = \begin{cases} 0 & \text{if } C \notin W, \\ 1 & \text{if } C \in W. \end{cases}$$

The Shapley–Shubik voting power index (1954) is based on the incremental or marginal value a party contributes to a coalition by joining it, weighted by the probability that the party will actually join it.

For example, let A, B, and C be three parties with 5, 45, and 50 seats each. A winning coalition needs at least 51 seats. Different sequences of parties define the construction of the ABC "grand" coalition, and according to the order which is followed, the contribution of single parties can be evaluated. The list of alternatives is as follows, with the corresponding contributions in brackets:

$$\begin{aligned}
&\text{AB}(\chi(\text{AB}) = 0) \text{ and then } C(\chi(\text{ABC}) - \chi(\text{AB}) = 1);\\
&\text{AC}(\chi(\text{AC}) = 1) \text{ and then } B(\chi(\text{ABC}) - \chi(\text{AC}) = 0);\\
&\text{BA}(\chi(\text{BA}) = 0) \text{ and then } C(\chi(\text{ABC}) - \chi(\text{BA}) = 1);\\
&\text{BC}(\chi(\text{BC}) = 1) \text{ and then } A(\chi(\text{ABC}) - \chi(\text{BC}) = 0);\\
&\text{CA}(\chi(\text{CA}) = 1) \text{ and then } B(\chi(\text{ABC}) - \chi(\text{BC}) = 0);\\
&\text{CB}(\chi(\text{CB}) = 1) \text{ and then } A(\chi(\text{ABC}) - \chi(\text{CB}) = 0);
\end{aligned}$$

Now suppose that these six alternatives have the same probability of occurring (i.e., 1/6). This means that the incremental value of party A is given by

$$\frac{1}{6}(0-0) + \frac{1}{6}(1-0) + \frac{1}{6}(0-0) + \frac{1}{6}(0-0) + \frac{1}{6}(0-0) + \frac{1}{6}(0-0) = \frac{1}{6},$$

since only in case 2, when A joins C, does this party increase the value of the coalition from 0 to 1. Similarly, the incremental value or expectation of party B is

$$\frac{1}{6}(0-0) + \frac{1}{6}(0-0) + \frac{1}{6}(0-0) + \frac{1}{6}(1-0) + \frac{1}{6}(0-0) + \frac{1}{6}(0-0) = \frac{1}{6}$$

and for party C it is

$$\frac{1}{6}(1-0) + \frac{1}{6}(0-0) + \frac{1}{6}(1-0) + \frac{1}{6}(0-0) + \frac{1}{6}(1-0) + \frac{1}{6}(1-0) = \frac{2}{3},$$

since B makes the coalition's value increase from 0 to 1 only in case 4 and C does in cases 1, 3, 4, and 6. The voting power vector (or Shapley value) of the game is, therefore, $(\frac{1}{6}, \frac{1}{6}, \frac{2}{3})$.

It can be noticed that the incremental value of each party is not a linear function of the seats they control since although A and B have the same voting power, they hold a very different amount of seats (A represents only 5% of the total, while B represents 45%). Another way to measure voting power is suggested by Banzhaf (1965), and it is based on the concept of critical defection. Given a winning coalition, say, C, the defection of party i is critical if the coalition $C - \{i\}$ is no longer a winner.

Banzhaf computes the power of a party in terms of the number of defections that are critical to the coalitions it can belong to, divided by the total number of critical defections, whatever the defecting party is. In the example mentioned above, it is easy to see that party C is a crucial actor in the following winning coalitions: ABC, AC, and BC, while party A is critical only for AC and party B is critical only in BC. In total, there are five distinct coalitions in which the defection of one of the parties is crucial, but only three of them are distinct coalitions. According to the Banzhaf index the voting power or party C is equal to 3/5, while that of party A and party B is equal to 1/5.

A more complex indicator has been proposed by Deegan and Packel (1980). In such case, only minimal winning coalitions are taken into account, and it is considered that they all have the same probability to occur. In addition, this index considers the fact that the parties involved in a coalition will equally divide the power within the coalition.

The Deegan and Packel voting power index for a party i is computed as follows:

$$\xi_i = \frac{1}{|C_m|} \sum_{C:C \in C_m, i \in C} \frac{\chi(C)}{|C|},$$

where C_m is the set of minimal winning coalitions.

Again, this index is not monotone with respect to the number of seats each party represents. Suppose there are 100 seats divided among five parties as follows: A has 35 seats, B has 20, and C, D, and E have 15. The set of minimal winning coalitions is

$$C_m = \{\text{AB, ACD, ACE, ADE, BCDE}\}.$$

Notice that in every coalition each party turns out to be critical by defection. Since the number of minimal winning coalitions is five, it can easily be shown that

$$\begin{aligned}
\xi_A &= \frac{1}{5}\left(\frac{1}{2}+\frac{1}{3}+\frac{1}{3}+\frac{1}{3}\right) = \frac{9}{30}, \\
\xi_B &= \frac{1}{5}\left(\frac{1}{2}+\frac{1}{4}\right) = \frac{3}{20}, \\
\xi_C &= \xi_D = \xi_E = \frac{1}{5}\left(\frac{1}{3}+\frac{1}{3}+\frac{1}{4}\right) = \frac{11}{60},
\end{aligned}$$

so that the voting power vector is given by

$$\xi = \left(\frac{18}{60}, \frac{9}{60}, \frac{11}{60}, \frac{11}{60}, \frac{11}{60}\right).$$

The voting power of party B, which holds 20 seats, is smaller than that of party C, D, or E although they only hold 15 seats.

The same occurs when the Holler index for voting power is adopted. Holler (1982) concentrates on minimal winning coalitions and measures the number of times a specific party i is a member of a minimal winning coalition, divided by the total number of times any party is a member of a minimal winning coalition. Therefore, the voting power attributed to a party i is given by

$$h_i = \sum_{C:C \in C_m i \in C} \chi(C) \frac{1}{\sum_j \sum_{C:C \in C_m j \in C} \chi(C)}.$$

For the example examined above, we get a voting power index according to Holler equal to $\boldsymbol{h} = \left(\frac{4}{15}, \frac{2}{15}, \frac{3}{15}, \frac{3}{15}, \frac{3}{15}\right)$.

To better understand why a party like B turns out to have less voting power than a party with a smaller number of seats, notice that a party with a small number of seats might be more likely to belong to a minimal winning coalition. In this way, a smaller party may be more important when the actors competing in Parliament are coalitions rather than single parties.

The voting power indexes described up to now (and others; see Gambarelli, 1994) are all based on the idea of measuring how crucial each party is in a coalition, although the set of coalitions taken into account can vary. Moreover, the size of the coalitions examined seems to be the principal factor, while the ideological or political distance between the parties in the coalition is never mentioned. Since political differences between parties are usually the first factor taken into account when building a coalition, we think it is reasonable to use a voting power index based on the subset of minimal winning coalitions whose political distance is lower than a given threshold, say, d. Henceforth, such a set will be denoted by C_{dM}. The choice of the threshold d strictly depends on the political context that is analyzed, but it should not be so restrictive that only the minimum distance coalitions are considered. In general d will depend on the number of competing parties, since the larger the number of actors on the scene, the more probable the need for wide coalitions, given that each party will tend to have a small number of seats. Consequently, the distance threshold must be wider to allow winning coalitions to be built. A reasonable threshold is broadly $\lceil \frac{n}{2} \rceil$, where n is the number of parties.

At this point a voting power index that takes political distance between parties into account could be referred to a given party i:

$$m_i = \frac{n_i w_i \sigma_i}{\sum_{j=1}^{n} n_j w_j \sigma_j},$$

where n_i is the number of distinct coalitions in the set C_{dM} to which party i belongs, w_i is the sum of the inverses of the number of parties that belong to the same coalitions as i, and σ_i is, as usual, the share of seats party i holds.

This index is proportional to the inverse of the number of parties belonging to the same coalition as party i, to the number of times i belongs to a coalition in C_{dM}, and to the number of seats it holds. This last feature implies that a distinction is made between parties that belong to the same number of coalitions and have the same number of other parties to bargain with but have a different number of seats, that is, a different weight within the coalitions.

Referring to the example described above, consider a Parliament with 100 seats distributed among five parties as follows: A holds 35 seats, and B, C, D, and E hold 20, 15, 15, and 15 seats, respectively. The distance threshold we consider is $d = \lceil \frac{n}{2} \rceil = 3$. If the parties, from A to E, are already arranged in a left-to-right order, the only coalitions belonging to the set C_{dM} where $d = 3$ are AB with 55 seats and distance equal to 1 and BCDE with 65 seats and distance

3. Then the voting power index corresponding to each party is

$$\begin{aligned} m_A &= \frac{1 \times 0.5 \times 35}{58.75} = 0.3, \\ m_B &= \frac{2 \times (0.5 + 0.25) \times 20}{58.75} = 0.51, \\ m_C &= m_D = m_E = \frac{1 \times 0.25 \times 15}{58.75} = 0.06. \end{aligned}$$

Again, the voting power index is not a monotone function of the seats, since the power attributed to party A (which is the largest in terms of seats) is significantly lower than that assigned to party B. In this specific case, the reason is that A cannot choose to join anyone else than B to form a winning coalition with a distance shorter than 3, while party B is in the middle of the left-to-right political distribution and can, therefore, rely on a wide span of alternative coalitions.

3.6 Government Stability

The great interest political scientists demonstrate for coalitions is mainly due to the fact that, in most cases, coalitions are needed so that a government can be established. In fact, in this analysis a government coincides with the set of those parties that have decided to join together in a winning coalition. Moreover, the analysis of governments in democratic regimes is concentrated on how long they last or how stable they are.

The historical approach to the analysis of government stability can be seen as an a posteriori examination of real-world cases in many different countries and times. It attempts to identify the factors or variables involved in creating instability, mainly by regression or other statistical tools. A widely cited work in this field is that of Herman and Taylor (1971), who study the relationship between government stability, measured in terms of how many days it has lasted, and the main variables characterizing a political system, such as the number of parties, the degree of fractionalization, etc. In particular, they show that Rae's fractionalization index tends to decrease when government stability increases and also when the larger number of parties is outside of the government. Moreover, many researchers tend to assume that the political or ideological distance between the parties in the government does not have a significant effect on its stability. This has lead to an important debate on how political distance should be measured [Warwick, 1979]. The left-to-right axis typically used to define the distance between parties should not be considered as the unique criterion, but other criteria, usually referring to the social features each party represents (such as ethnic, religious, or social minorities), should also be taken into account.

Another approach to the analysis of government stability is based on forecasting or measuring the expectation for a stable government which can be related to different political scenarios. That is, given an electoral outcome, the different coalitions that can allow a government to be established must be evaluated. Intuition leads us to suppose that the larger the number of winning coalitions, the smaller the degree of stability that can be expected.

As we have already mentioned, to receive the favor of the majority of the voters, a government can be established if it represents a winning coalition, that is, more than 50% of the seats. Since the political distance between parties makes some winning coalitions highly improbable, from now on we will only refer to the subset C_{dM}, where d is a given distance threshold.

The expectation of government stability can be measured by an indicator, denoted by ES, which takes into account the following requirements:

- ES should increase when the number of seats each coalition holds tends to increase;
- ES should increase when there exist "strong" parties, that is, parties that frequently appear in the set of winning minimum distance coalitions;
- ES should decrease when the number of coalitions in the set increases;
- ES should decrease when the number of parties needed to build a winning minimum distance coalition increases.

Such an index can be computed as follows:

$$ES = \frac{1}{|C_{dM}|} \sum_{C \in C_{dM}} \frac{w(C)\pi(C)}{|C|},$$

where $\pi(C)$ is the percentage of total seats that coalition $C \in C_{dM}$ holds and $w(C)$ is a weight associated with $C \in C_{dM}$, calculated on the basis of the number of parties it contains, as shown in the following example.

Suppose there are 100 seats distributed among five parties already ordered from left-to-right as follows: A holds 35 seats, B holds 20 seats, and C, D, and E hold 15 seats each. The set of minimum distance winning coalitions (with $d = 3$) is given by {AB(55), BCDE(65)}. Let us first count the number of times each party appears in one of these coalitions and call this index $IC(i)$ for party i:

$$IC(\text{A}) = 1,\ IC(\text{B}) = 2,\ IC(\text{C}) = 1,\ IC(\text{D}) = 1,\ IC(\text{E}) = 1.$$

Summing up the values of IC corresponding to each party, we can give a value for each coalition and, namely, $IC(\text{AB}) = 3$ while $IC(\text{BCDE}) = 5$. The weight assigned to each coalition corresponds to

$$w(C) = \frac{IC(C)}{\sum_{C' \in C_{dM}} IC(C')}.$$

In our case w (AB) = 0.375 and w(BCDE) = 0.625. Therefore, ES is equal to

$$ES = \frac{1}{2}\left(\frac{0.375 \times 0.55}{2} + \frac{0.625 \times 0.65}{4}\right) = 0.102.$$

ES will be equal to its maximum value ($= 1$) when all of the seats available are assigned to the same party, so that the only possible coalition is made by the party itself. On the other hand, it tends to 0 when the number of minimum distance winning coalitions C_{dM} tends to infinity.

Let us examine a few examples to better understand how the expectation for government stability and the voting power of each party interact.

First suppose there are 100 seats distributed among five parties ordered on the left-to-right axis as follows: 10 seats for party A, 15 for B, 30 for both C and D, and 15 for E. Now, if the distance threshold d is equal to 3, the set of minimum distance winning coalitions is $C_{dM} = \{\text{ABC}(55), \text{CD}(60)\}$. The weight assigned to these coalitions is 0.571 for ABC and 0.429 for CD, respectively. It can be shown easily that the expectation for government stability is equal to

$$ES = \frac{1}{2}\left(\frac{0.571 \times 0.55}{3} + \frac{0.429 \times 0.6}{2}\right) = 0.117.$$

In the meantime, we can compute the voting power vector where political distance between parties is considered (the components m_i of such vector have been defined in section 3.5) and obtain $\boldsymbol{m} = (0.04, 0.07, 0.68, 0.2, 0)$. On the other hand, if the seats are distributed so that A, B, C, D, and E control 18, 15, 30, 22, and 15 seats, respectively, the set of minimum distance winning coalitions is $C_{dM} = \{\text{ABC}(63), \text{CD}(52)\}$, $ES = 0.116$, and $\boldsymbol{m} = (0.08, 0.07, 0.69, 0.15, 0)$. Finally, if the seat distribution is 10, 20, 25, 30, and 15 seats to party A, B, C, D, and E, respectively, the set of minimum distance winning coalitions is $C_{dM} = \{\text{ABC}(55), \text{CD}(55)\}$, so that $ES = 0.111$ and $\boldsymbol{m} = (0.05, 0.1, 0.62, 0.22, 0)$. Although these three cases are very similar, since the set of minimum distance winning coalitions is the same, stable government expectation tends to slightly decrease because the relative strength of the parties within each coalition changes. The highest expectation value is registered when the first example is considered, since in this case there is a coalition with a higher amount of total seats (62) but only two parties. This makes the bargaining process between the parties easier and, therefore, represents a guarantee for the stability of the government.

In the third case the seat distribution is less concentrated than in the others. The consequence is that the winning coalitions control a smaller amount of seats. This is why ES tends to assume a lower value. The fact that $\boldsymbol{m}$ is less concentrated than in the other two cases also shows that there are no "strong" parties, making the alternatives more numerous and thus suggesting less stability.

These examples show that the ES index is sensitive to the strength of the parties, that is, to the number of seats each party and each coalition controls. It also depends on the number of minimum distance winning coalitions. In fact, if the same number of available seats were distributed among five parties (A, B, C, D, and E) ordered from left to right with 20, 24, 20, 16, and 20 seats, respectively, the number of minimum distance winning coalitions would be three ($C_{dM} = \{\text{ABC}(64), \text{BCD}(60), \text{CDE}(56)\}$), $ES = 0.067$ and $\boldsymbol{m} = (0.05, 0.25, 0.47, 0.17, 0.05)$. But, on the other hand, if there are six parties

(A, B, C, D, E, and F) which control 40, 20, 10, 20, 5, and 5 seats, respectively, then there is only one minimum distance winning coalition, AB, with 60 seats. Then $ES = 0.3$ and $\boldsymbol{m} = (0.67, 0.33, 0, 0, 0, 0)$.

In the first case, the seats are more or less equally shared between the different parties, so that many coalitions are possible and they are all of the same size (that is, they all contain three parties). In this kind of situation it is easy to imagine a strong variability of the political scenario, and we cannot expect a stable government. On the other hand, when the seats tend to be concentrated in the hands of a small number of parties, a unique coalition is possible and ES tends to assume high values.

Part II

Designing Electoral Systems

All animals are equal but some animals are more equal than others.

George Orwell, *Animal Farm*

Democratic representation systems require the *fair* selection of a small number of individuals to represent a much larger number of citizens. The electoral system is the constitutional tool used to solve this problem by transforming citizens' votes into legislature seats. But, the fact is, there exists a wide variety of electoral procedures to choose from. History, political debate, and national heritage have led to the adoption of a variety of different systems throughout the last two centuries and, today, electoral reform seems to concern new-born democracies as well as the old ones. Counting the number of distinct electoral systems that are technically or historically relevant is a matter of no importance, but it might be surprising to discover that more than 300 different proportional electoral systems had already been classified in 1951 [Hermens, 1951]. This shows how difficult it can be to find a unique solution to the representation problem. The classification and the coding system described in the first part of this book show that a huge number of different electoral procedures can be created by a simple combination of the basic variables such as the kind of ballot, the size of the districts, and the kind of electoral formula. However, arbitrary combinations of these variables do not guarantee that the method adopted will always find a seat allocation and, in any case, we want the seat allocation found to satisfy some basic properties in order to be fair.

As shown in the following chapters, designing new electoral systems is just as hard as having to choose among the many existing procedures. Cultural, historical, and, maybe, accidental factors are involved, but the fact is that the choice of an electoral system cannot be considered impartial or unbiased. Voting in itself is not a guarantee of fairness since the procedure adopted can make a radical difference in the voting outcome.

This is the major topic of Part II, which concerns *electoral engineering* and in particular the design, the analysis, and the selection of adequate electoral procedures. Our attention will be focused on the procedures used to transform votes into seats (electoral formulas) which constitute the core of every electoral system.

In Chapter 4 we review the traditional methodology used to evaluate voting procedures and we discuss the impossibility of designing an electoral system which never leads to contradiction with respect to expectations of common sense. A more extensive analysis of the main properties associated with electoral formulas is given in Chapter 5. These properties are nothing more than formal statements that represent the common sense described above and constitute the basic evaluation criteria for electoral procedures in the traditional, axiomatic approach. A minimal list of reasonable properties that every electoral formula should satisfy is then considered. Unfortunately, just as stated in the "electoral version" of Arrow's impossibility theorem, there exists no *perfect* electoral system (that is, no system that simultaneously satisfies such a list of properties). The choice of an appropriate electoral system is thus reduced to the choice of a feasible combination of the properties that are considered fundamental.

In Chapters 6 and 7 we present an integer optimization approach to the analysis and design of proportional seat assignment methods. In fact, electoral formulas often are not mathematical "formulas" in the traditional sense, but they operate more like algorithms that receive an input (the vote distribution) and generate an output (a particular seat assignment) through a sequence of elementary operations. In our view, each electoral formula is a specific algorithm designed to minimize a particular cost function. In particular, proportional formulas are efficient algorithms used to solve an integer optimization problem with specific objective functions, some of which are illustrated in detail. The objective functions, which are different according to each formula, represent different proportionality indexes. They can be viewed as the "hidden" criteria on which each formula is based. We think that this kind of approach can give a theoretical background to the well-known proportional formulas. In fact, this will allow us to point out the weakness of common methodologies in the analysis of proportional systems and to introduce the concept of Schur-convex functions as a relevant class of disproportionality indexes. Schur convexity had a key role at the beginning of this century, when the pioneers of *welfare economics* were concerned with comparing different income or wealth distributions. We use it in the electoral context to measure the inequality of the seat-to-vote ratios in the legislature apportionment.

In this context, the choice of an appropriate electoral formula can be translated into the definition of an appropriate measure of fairness (not necessarily restricted to the ones that will be mentioned) and the design of an algorithm which optimizes it. This approach will at least guarantee that the chosen electoral procedure be consistent with our expectations.

In Chapter 8 we will return to considering the variety of electoral systems by analyzing a special class of procedures obtained as a combination of both majority and proportional methods. The criteria described in Chapter 5 cannot be adopted to analyze these methods, which are generally called *mixed* or *hybrid* electoral systems. In fact the idea underlying these procedures is that of finding a desired balance between the effects usually attributed to these two opposite classes of formulas. Mixed systems seek the "best of both worlds" trying to ensure the establishment of a stable government and a high level of representation of the preferences pronounced by the voting body. We will therefore be particularly interested in analyzing the impact of a variation of the share of seats assigned by a majority formula in the mixed system on the electoral campaign strategies and on the electoral outcomes.

The methodology presented in this part of the book will not tell the reader how to design or to choose the "best" electoral system; however, it provides the reader with a useful toolkit to explain how votes can be transformed into seats and to understand why, although all parties are equal, "some are more equal than others."

Chapter 4

No Electoral System Is Perfect

The quantitative analysis of electoral formulas is traditionally based on two different methodological approaches:

- the *axiomatic approach*, based on the definition of the properties that characterize single methods and different classes of methods; these properties are used to advocate the legitimacy and the fairness of the methods themselves;
- the *functional approach*, based on the measurement of the effects of the different electoral formulas on the political system as a whole, in terms of a set of performance indicators such as representation indexes, the number of political parties, the fractionalization of vote and seat shares, and government stability, but also the cost of elections, the participation to vote, and the sensitivity to change in vote distributions.

The first big step toward the development of the functional approach must be attributed to *The Political Consequences of Electoral Laws*, written by Douglas Rae in 1967. The electoral system is described as a political tool which acts on the whole political system, on representation, and on the party structure. In this context, electoral reform is considered more as a way to modify an undesired structure of the political system than a way to respond to the evolution of the concept of democracy. Therefore, electoral formulas are classified on the basis of their consequences on the number of political parties, on the possibility of creating majority coalitions and stable governments, and many other criteria of this kind [Blais, 1988; Lijphart, 1985a, 1985b; Loosemore and Hanby, 1984; Nilson, 1974].

Nevertheless, comparing electoral formulas on the basis of their political effects can often be inadequate and in some cases even misleading. A superficial analysis equates "stable government" and "majority formulas," as well as "representation" and "proportional formulas." But the consequences of an electoral system are not solely determined by the procedure used to translate votes into seats. A fundamental role is played by the other variables of the system, such as the kind of ballot and the size of the districts. The consequences, say on representation, of specific combinations of district structures and proportional formulas can be the same as those of some majority systems, and vice versa. Another fact that we would like to point out is that most of the results in this field come from statistical analysis. This might not always be very convincing because it proceeds on a heterogeneous ground, comparing situations that differ in time and space and for which no controlled experiment can be performed. In some cases it is impossible to separate the effective and the spurious correlation observed between variables. Ultimately, we must not ignore the effect of the social and economic development and the historical and cultural background on the origin and the structure of the political parties and the government.

This does not mean that the criteria adopted to select a specific electoral formula must exclude the will to modify the political structure of a country or to obtain well-defined effects. Electoral systems do act on the political system, emphasizing or reducing the patterns displayed in the vote distribution.

The rest of this chapter is dedicated to the axiomatic approach, which better reflects our interest in the technical features of electoral procedures and their inherent properties.

When we talk about inherent properties, we mean those consequences of the electoral formula that do not depend on any other factor, such as the specific vote distribution or the structure of the political system. Identifying such consequences is the first step toward the definition of the properties that we require a *good* electoral system to bear. Unfortunately, the axiomatic approach cannot always justify the choice of a specific electoral system because of the many paradoxes and tricky contradictions every method seems to produce. In a slightly different context, the interest for such a topic has been alive for several decades. In fact, when referred to the theory of social choice, the whole debate culminates with Arrow's famous *impossibility theorem* (1963). We will therefore point out how an axiomatic approach leads to an equivalent electoral impossibility theorem. The purpose is to prove that no electoral system is perfect, nor it is possible to design a perfect system. The following paragraphs give an idea of the difficulties that may arise by providing several examples of the many paradoxes that have been identified in this area.

4.1 Famous Paradoxes

Since the beginning, paradoxes have accompanied the development of social choice theory. Many *voting paradoxes* have fascinated famous researchers and have animated fierce intellectual debates.

The Condorcet Paradox

If there is a unique seat to assign and more than two candidates, we would expect any electoral formula to choose the candidate who is preferred by the majority of the voters. But when no candidate obtains more than 50% of the total votes it may not be so clear who should win the seat. An idea could be that of considering the most preferred candidate as the one who always obtains the majority of votes in each face-to-face comparison with the other candidates. Such a candidate is known as the *Condorcet winner* and it seems natural that he or she should win the elections, provided that it exists. Nevertheless, there are cases in which a candidate receives a strict majority of votes in every pairwise competition but loses the election. In fact, this can occur even with very common voting procedures such as the first-past-the-post method. Suppose there are three candidates (A, B, and C) and that the voters can be divided into different groups with respect to their preferences. Each voter orders the candidates according to his/her personal preference and votes for his/her first choice.

In the example given in Table 4.1, the winner is candidate A when the first-past-the-post method is adopted. But a careful examination of the table shows that if candidate B is compared with candidate A without considering C, then B would receive a larger number of votes (700 against 600), so as to say that the majority of the voters actually prefer B to A. Moreover, B would continue to be the most preferred candidate in a pairwise competition with C (900 votes for B against 400 for C). In such a situation, B seems to be the candidate preferred by the vast majority of voters, but the first-past-the-post method chooses A.

Table 4.1. *Example* 1.

	First choice	*Second choice*	*Third choice*
600 voters	A	B	C
400 voters	C	B	A
300 voters	B	C	A

The Alabama Paradox and the New States Paradox

The paradoxes we are now going to discuss arise in the context of seat apportionment. In Chapter 2 we have already pointed out the similarity between the problems of finding a fair apportionment of seats to districts and a fair apportionment of seats to lists or cartels. This will be examined further in the next chapter.

Two of the most famous paradoxes, the *Alabama paradox* and the *paradox of new states*, are strictly related to contradictions that have occurred in the apportionment of seats to the different states in the United States. Technically the same kind of inconvenience can occur while assigning seats to parties, but historically the major debates have been concerned with the allocation of seats to geographical regions or states.

In 1881 Largest Remainders was sharply criticized and then abandoned by the U.S. Congress because it produced an unpleasant effect on the seat distribution. The number of seats in Congress increased from 299 to 300, but only 7 seats were assigned to the state of Alabama while it would have received 8 seats if the total number of seats in Congress had not changed.

In 1907, another peculiar event left its signature in apportionment history. That year Oklahoma became a new state and 5 seats were reserved for its representation. Thus, the total representation increased from 386 to 391 seats, 5 of which were assigned to Oklahoma on the basis of its population. Although the population of the old states had not changed, the entry of a new state created some side effects on the overall seat distribution. In particular, without any apparent reason, one seat that was previously assigned to New York went to increase those assigned to Maine. New York passed from 38 to 37 seats, while Maine passed from 3 to 4 seats, thus acquiring extra power in Congress at the expense of New York.

The contradiction underlying these two paradoxes is similar and has to do with monotonicity: an increase in the number of available seats causes some party (or region or electoral district) to receive a smaller number of seats with respect to the number it would have obtained if the seat total had remained unchanged. Somehow, the probability of receiving a seat should be higher simply because more seats are available. The problem is that someone is damaged in the final allocation in spite of the fact that the same vote (or population) distribution is considered. Both paradoxes arise from the fact that Largest Remainders is not *house monotone*.

We would like to point out that these paradoxes cannot be considered as a mere academic example or as a rare historical event. Phenomena of this kind have occurred in many situations, and their political consequences are often more than a simple inconvenience. In fact, most debates on proportional formulas have been fought over the designation of a single seat. Just recently, in 1991, two separate cases presented by Montana and Massachusetts challenged in court the constitutionality of an apportionment method. The method under discussion was Equal Proportions, which had been used for over 50 years. Both Montana and Massachusetts proposed alternative methods after being penalized by Equal Proportions on the basis of the 1990 census, but in 1992 the Supreme Court upheld the constitutionality of the current method [Ernst, 1994].

4.2 Other Electoral and Voting Paradoxes

Could it be convenient to vote for a candidate or a party that is not our first choice? In other words, could a party ask its supporters to vote for an adversary to guarantee its own victory? Can a party with the largest number of votes actually lose the elections? Can two parties that decide to share their power actually lose it? In the following sections some examples of these paradoxes are shown.

The Sincerity Paradox

In some cases, by decreasing the number of votes it receives, a party can increase the number of seats it obtains under certain assignment methods. This typically happens with the double ballot method, where voters who vote sincerely revealing their true preference might discover that by supporting their favorite party they have actually damaged it. In this case, we are dealing with the *sincerity paradox*.

Suppose there are three parties, A, B, and C, and 1700 voters, and consider the preference profile shown in Table 4.2.

Table 4.2. *Example* 2.

	First choice	*Second choice*	*Third choice*
800 voters	A	B	C
500 voters	C	A	B
400 voters	B	C	A

If each elector votes "sincerely," that is, supporting the party that he prefers to all others (first choice), A and C will be competing in the second ballot, since no party receives more than 50% of the vote total. This forces the 400 supporters of party B to choose between A and C, and again if they vote for their second best choice, party C will be the overall winner of the election.

If party A asks 200 of its supporters to vote for party B in the first ballot, the two parties in the second ballot would be A and B, with 600 votes each. This time the supporters of party C will have to decide between A and B. Following the previous argument, this will lead A to be the winner with 1300 votes against 400. So, in this case, the only way for a group of voters to allow their favorite party to win is to vote for an adversary.

The Strongest Party Paradox

A very frequent paradox arising in the context of majority formulas such as the first-past-the-post method is *the paradox of the strongest party*. It occurs when the party that collects the largest number of votes in a multidistrict region (and therefore seems to deserve a consistent amount of seats) is damaged by its opponents in each single-member district to the point that it does not receive any seats at all. Such a party is overall the most powerful in terms of votes, but it does not obtain any representation.

For example, consider 5 seats that must be assigned to 4 parties with the first-past-the-post method. The territory is divided into 5 single-member districts with the vote distribution (in thousands of votes) given in Table 4.3.

Table 4.3. *Example* 3.

Party	*District* 1	*District* 2	*District* 3	*District* 4	*District* 5	Total votes
A	**40**	14	10	7	**35**	106
B	14	15	**42**	10	15	96
C	38	35	40	41	33	187
D	8	**36**	8	**42**	17	111
	100	100	100	100	100	500

Winners in the single-member districts are those who receive the majority of votes in those districts and they are marked in bold print. Thus, the final seat assignment is the following: 2 seats to party A, 1 to party B, and 2 to party D. No seat has been assigned to party C, although this party is the largest in terms of votes. In total, party C obtains 76% more votes than party D but no representation. The problem is that its supporters are very homogeneously spread out over the whole territory so that it loses by just a handful of votes in each single-member district.

The Coalition Paradox

It is natural to suppose that if two parties merge into one, sharing their votes, the new party obtains at least the sum of the seats the two parties would have obtained separately. But, with some common electoral methods, such as Largest Remainders, party coalitions can sometime lose part of the seats they would have won if they had remained separate.

Consider the example in Table 4.4, where there are three seats to assign to three parties.

Table 4.4. *Example* 4.

Party	*Votes*
A	14,000
B	7,000
C	6,000
Total	27,000

Under the Largest Remainders method each party will receive exactly one seat since their exact quotas are 1.55 for A, 0.78 for B, and 0.67 for C. If B and

C decide to form a coalition against A and join together in a single party with a total of 13,000 votes, the exact quota would change to 1.55 for A and 1.45 for (B + C). But, in such case, party A would receive two seats and party (B + C) only one, although separately B and C would have obtained one seat each.

In such a case, a coalition will make both parties weaker in terms of legislative power. Methods that are included in this paradox do not encourage coalitions, but on the contrary they encourage parties to split up into smaller ones only for electoral purposes.[3] These methods do not satisfy the *superadditive property*.

4.3 An Electoral Version of Arrow's Impossibility Theorem

No electoral system is perfect, because in many cases there simply does not exist a solution satisfying all the properties we would naturally require. The purpose of the following simple example is to show that it is not possible to design a vote-to-seat transformation procedure which will never be criticized on the basis of an axiomatic approach (see also Scozzafava, 1990; Bernardi and Menghini, 1990). This is nothing more than an extension of Kenneth Arrow's impossibility theorem for social choice functions. In fact, an electoral system maps individual choice functions into collective or social choice functions.

Suppose we want to assign 3 seats to 4 parties in a district of 100,000 electors with the vote distribution given in Table 4.5.

Table 4.5. *Example* 5.

Party	*Votes*
A	47,000
B	24,000
C	17,000
D	12,000
Total	100,000

The fairness of a seat assignment can be tested by checking whether it satisfies a set of conditions which prevent paradoxes to occur. The fulfillment of such conditions can be thought of as a first certificate of the fairness of the procedure. Actually, we want the seat assignment to satisfy the following properties:

(1) if the number of votes party A obtains is greater than the number of votes party B obtains, then the number of seats assigned to party A must be no

[3] In this context the effects of coalitions and schisms on the fluctuation or shift in the voting preferences are not considered.

less than the number of seats assigned to party B:

$$\text{votes(A)} > \text{votes(B)} \Rightarrow \text{seats(A)} \geq \text{seats(B)};$$

(2) a party, say, A, receives a strict majority of seats if and only if it has obtained a strict majority of votes:

$$\text{seats(A)} > 50\% \text{ of seats} \Leftrightarrow \text{votes(A)} > 50\% \text{ of votes};$$

(3) the sum of the seats assigned to a coalition of two parties considered as a whole cannot be less than the sum of seats they would have obtained separately:

$$\text{seats (A + B)} \geq \text{seats(A)} + \text{seats(B)}.$$

The properties stated above are generally known as *vote monotonicity* (1), *the majority principle* (2), and *superadditivity* (3).

Now, since the parties are 4 and the seats are 3, the number of different seat assignments is equal to the total number of ways to write the integer 3 as the sum of 4 nonnegative integers (order does matter here):

$$\binom{4+3-1}{3} = \frac{6!}{3!3!} = 20.$$

These 20 distinct seat assignments are listed in Table 4.6.

It is easy to check that the only seat assignments that satisfy monotonicity (property 1) are the ones in bold print. Nevertheless, assignment 1 does not respect the majority principle (property 2) because it gives all three seats to party A, which has obtained only 47% of the votes. The same holds for assignment 5, where party A receives 66% of the total legislative power having obtained fewer than the majority of votes. Therefore, the only seat assignment left is 17. But this assignment will satisfy superadditivity (property 3) only if it violates the majority principle (property 2). In fact, an electoral procedure which yields assignment 17 would also have to assign the coalition (B+C) a number of seats at least equal to the sum of seats they already have received separately. So assume that seats(B+C) = seats(B) + seats(C) = 1 + 1 = 2. In such an instance, property 2 is not satisfied because the new party (B+C) receives a strict majority of seats even if it holds only 41.1% of the total votes.

The theory of social choice has defined many different fairness criteria stating how an ideal electoral procedure should behave. When a seat assignment procedure fails to satisfy a fairness criterion, it falls into a paradox and produces results that may be considered unfair to the electorate. This would be contrary to the nature of voting itself, which is that of finding a fair compromise to synthesize the divergent opinions of the members of a group. Unfortunately, "impossibility theorems" prove that no social choice function (and electoral systems give rise to particular social choice functions) can simultaneously satisfy more than very few fairness criteria at the same time. Thus, there seems to be

Table 4.6. *Example* 6.

	A	B	C	D
***case* 1**	**3**	**0**	**0**	**0**
case 2	0	3	0	0
case 3	0	0	3	0
case 4	0	0	0	3
***case* 5**	**2**	**1**	**0**	**0**
case 6	2	0	1	0
case 7	2	0	0	1
case 8	1	2	0	0
case 9	1	0	2	0
case 10	1	0	0	2
case 11	0	2	1	0
case 12	0	2	0	1
case 13	0	1	2	0
case 14	0	1	0	2
case 15	0	0	2	1
case 16	0	0	1	2
***case* 17**	**1**	**1**	**1**	**0**
case 18	1	1	0	1
case 19	1	0	1	1
case 20	1	1	0	1

no escape: whichever the seat-assignment procedure we use, there is a risk of falling into contradiction.

Most attempts to overcome Arrow-like theorems are based on arguing that the fairness criteria that are violated by a given method are not reasonable or fundamental criteria or on showing that there are trends in voting behavior that naturally prevent paradox. Another common attitude is that of discarding the relevance of the contradictions proving that they are determined by theoretical situations but that they rarely occur in practice. Many of the cases discussed above are, in fact, purposely made examples. Their importance, therefore, is strictly related to the probability they occur in real electoral situations. Somehow this probability is generally considered too low to be a threat, although the literature is very poor in this field. The only paradox that seems thoroughly examined in terms of probability is that of *cyclical majorities* [DeMeyer and Plott, 1970]. Since no seat-assignment procedure is immune to paradox, in the sense that it cannot simultaneously satisfy more than very few fairness requirements, why not just accept that unfairness can occur? In fact, our major concern is not unfairness in itself, but rather the fact that the method can be systematically biased in favor of some party or candidate and the fact that someone might exploit the potential unfairness to control the voting process and manipulate the results. Given that fairness criteria are not sufficient to protect the integrity of an electoral scheme, the problem is to determine, in practice, which schemes are harder to control than the others. This point of view provides another interesting criterion for the evaluation of electoral procedures based on computational complexity analysis. Procedures that are "hard" to manipulate are those who typically require unreasonable amounts of computer time (years or even centuries) to be manipulated, even if the controller has a complete knowledge of the preferences of all voters [Bartholdi, Tovey, and Trick, 1989].

In our opinion, arguments based on the probability that paradoxes occur or based on the complexity of manipulation are much more interesting on operational grounds than the many attempts to disable Arrow-like theorems by criticizing the choice of the criteria used to formalize the idea of fairness. On the other hand, a careful analysis of the properties underlying the different methods is necessary to understand their behavior and to identify the context in which they can be successfully applied. Politicians should argue the merits of electoral formulas comparing them on the basis of the properties which are considered the more appropriate, but they must not forget that in some cases contradictions do occur.

Chapter 5

Basic Properties for Electoral Formulas

Throughout the literature, a large number of different properties and axioms are taken into account in an attempt to find the minimal requirements for a seat allocation method to be fair. The axioms that are described below are thought of as properties that everyone would expect an "ideal" electoral formula to satisfy. However, as the many examples illustrated in the previous chapter prove, such an ideal system does not exist and, as this chapter points out, it cannot even be designed.

The set of properties a given formula satisfies is a useful way to characterize its technical virtues and vices. The relations between different properties are crucial in the design of seat allocation methods because they embody the conflict existing between different concepts of fair representation.

In order to analyze the electoral formula in itself, we must restrict our attention to a single district. For majoritarian methods this means that only one seat is assigned (single-member districts), while for proportional methods S seats are assigned (where $S > 1$). The consequence is that a direct comparison between majoritarian and proportional methods in terms of basic properties cannot be made in practice. The purpose underlying these two classes of methods—that is, democratic representation—is the same at national level but very different at district level, so they must be evaluated separately. Moreover, when majoritarian methods are considered, the votes are cast for a candidate while they are cast for a party when proportional methods are taken into account.

A general property anybody would require an electoral procedure to satisfy is symmetry. *Symmetry* or *anonymity* requires the seat assignment to depend only on the number of votes each party has obtained and not on the order of candidates or parties, their symbol, or label. Surprisingly enough some commonly used voting procedures do not even satisfy this property. An example is the *amendment procedure*, which consists of voting sequentially for pairs of alternative candidates just like in a tournament. The loser of the bilateral confrontation is eliminated, while the winner must compete with the next candidate, until all candidates have been considered. The outcome of this kind of method is strongly dependent on the order in which the candidates are brought into the pairwise comparison and, therefore, the procedure is affected by agenda control. Luckily, such a procedure would never be used in the context of Parliament elections, although it is commonly used in committees or councils (and even in Parliaments) in several decision-making processes.

5.1 Majoritarian Methods

From the start, social choice theory has been concerned with procedures designed to select a collectively best alternative (or subset) on the basis of individual preference rankings of the alternatives (or subsets of alternatives). The theoretical analysis of majoritarian electoral systems has developed within this framework benefiting from the solid axiomatic foundations which had been built for more general social choice functions. Therefore, the language and the approach adopted in the study of majoritarian methods is nearly totally derived from such a common background in research. The literature on the properties of social choice functions is extremely rich and enlightened by the work of famous mathematicians, economists, and social scientists, including several Nobel Prizes, but specific applications to the evaluation of electoral systems represent a minority share [Fishburn, 1983; Richelson, 1978; Merrill, 1984]. A good collection of results in this area can be found in Nurmi (1987). The following section should be considered a brief introduction to the subject, restricted to the concepts that are of interest in our case. Besides first-past-the-post and double-ballot, the most common majoritarian methods, recall the *alternative transferable vote system* described in Chapter 2. The voters rank all the candidates according to their preferences. In the single-member district case, the seat is automatically assigned to the candidate who is ranked first according to more than 50% of the voting body. If such a candidate does not exist, the candidate with the fewest first ranks is eliminated from further consideration and the other alternatives are pushed up in the ranks. Again, the new first ranked candidate is considered and the elimination procedure repeated until one candidate is ranked first by more than 50% of the voters according to the new preference orders.

Another system is the *approval voting* procedure [Brams and Fishburn, 1978], which can be considered as an extension of the first-past-the-post method, where voters are allowed to cast more than one vote at the same time following the spirit of "one man, k votes." That is, instead of asking voters only their most preferred candidate, they can vote for as many candidates as they wish without ordering them on a preference list but just giving them either one vote or no votes. The candidate with the largest number of total votes wins. The idea is that it is too difficult for the voters to give a complete ranking of a large set of candidates and, therefore, it is more reasonable to simply ask them which of the candidates they consider acceptable.

The properties related to majoritarian methods need each voter's whole list of preferences to be examined, whether it is fully considered (such as in approval voting and in the single transferable vote), partially considered (such as in the double ballot), or only the first choice is considered (such as in the first-past-the-post). This is in fact what has been done in the examples shown in Chapter 4.

Among the many properties that can be taken into account, the most widely cited in the literature of social choice theory are without any doubt the Condorcet principles, the Pareto principle, the weak axiom of revealed preferences, and path independence. The Condorcet principles are related to the transitivity of the collective choice relation; the Pareto principle defends unanimous preferences, while the other two properties are important to assess the robustness of the method when groups and subgroups of candidates are considered separately and to guarantee that the voting procedure is immune to manipulation on the set of candidates. In fact, it is reasonable to suppose that if a candidate wins against the others considered all together, he or she also wins against the others if they are considered in separate subgroups.

The Condorcet Principles

A *Condorcet winner* is an alternative that defeats any other candidate by a simple majority in pairwise comparisons. Such a candidate does not necessarily exist, but when it does, it seems reasonable to expect that it actually wins, whatever the procedure adopted. Electoral procedures that do not refer to the complete preference profile of voters to determine the final winner cannot assure that the Condorcet winner, even when it does exist, is actually chosen. In fact, this is what happens for the first-past-the-post method in the example shown in Table 4.1. Several other methods, including procedures that do consider complete rankings of the candidates, such as the single transferable vote, fail to satisfy the Condorcet winner principle; however, what is even more serious is that some methods, such as the first-past-the-post, can end up by choosing a candidate who would be defeated in pairwise comparisons by *each* of the other alternatives, that is, a *Condorcet loser*. A closer inspection of Table 4.1 shows that this is the case. If the fact that the seat is not assigned to the Condorcet winner is considered a flaw, a method that might assign it to the Condorcet loser is even more questionable.

Monotonicity

Single electors will find monotonicity a basic requirement for seat allocation methods since it guarantees that their vote actually supports their favorite candidate or party. A method is *monotone* if the number of seats assigned to a party does not decrease when the number of its supporters grows:

$$\forall i: \quad v_i' > v_i \Rightarrow s_i' \geq s_i.$$

The example in Table 4.2 shows that the double ballot method is not monotone.

The Pareto Principle

Consider two candidates, say, x and y. The *Pareto principle*[4] states that if all the voters prefer candidate x to candidate y, then y must not be chosen. Indeed, one would certainly agree that there is something wrong in a voting procedure that lets y be the winner even though everyone prefers x.

The Weak Axiom of Revealed Preferences (WARP)

Let X be a set of candidates. The *weak axiom of revealed preferences* requires that the following two conditions hold:

(1) if x^* is a winner in the set of candidates X, it must still be a winner in every subset of candidates $X' \subseteq X$ such that $x^* \in X'$;

(2) if there are ties between candidates belonging to the same subset of $X' \subseteq X$, the final set of winners in X must either include them all or exclude them all.

The first requirement of this property, which is also called *strict heritage*, sounds compelling: it basically states that a winner must remain such whatever subset of candidates (it belongs to) the procedure is applied to.

When a method satisfies the WARP, we will end up with the same candidate (or subset) both by applying the method to the set X and then taking the intersection between all winners in X and the subset X' and by directly applying the procedure to X' (under the assumption that the intersection between all winners in X and the subset of candidates X' is not empty).

Figure 5.1 shows a case in which the WARP is satisfied. The subset of winners in X' corresponds to the intersection between the winners in the larger set X and the set X' itself.

A method satisfies the WARP if this holds for any subset $X' \subseteq X$.

Fig. 5.1. *WARP is satisfied.*

Figures 5.2 (a) and (b) show different types of violation of the WARP. In particular, the first case violates condition 1 since there are winners in X who are not winners when X' is considered on its own. The second case violates condition 2 since there are ties in X', but a subset of the winners in X' is excluded when the winners in X are considered.

[4]The property we refer to is more precisely the weak Pareto principle. Many extensions and variations of this principle exist, but they will not be discussed here (see Nurmi, 1987).

(a) (b)

Fig. 5.2. (a) *Condition* 1 *of WARP is violated;* (b) *condition* 2 *of WARP is violated.*

This axiom is generally requested to guarantee the voting method be immune to manipulation on the set of candidates and, in particular, to the addition of "dummy" candidates to the election. Dummies are candidates who are introduced among the possible alternatives to split up the vote. Manipulation arises when such candidates are added with the intention of diverting a large amount of votes away from a given "real" candidate to indirectly assist some other specific candidate, although the dummy candidates will never win themselves. Let X' be the set of "real" candidates and X the enlarged set, including the dummy candidates. If a candidate x^* is a winner in X' and each voter's preference profile does not change, the set of final winners, when all of X is considered, must still contain x^*. In other words, the voting procedure must select the same winners and losers in X' both when X' is considered on its own and when the enlarged set X is considered. This will guarantee that the outcome is not affected by introducing dummy or straw-man candidates to the agenda, because no candidate who is a loser in X' will become a winner when X is considered and vice versa.

Path Independence

This principle claims that the final outcome of the voting procedure must be independent from the "path" adopted to select the winners. A procedure can be based on more than one selection phase and the final result could be affected by the particular order in which the candidates are compared. Again, the idea underlying path independence is that manipulation on the set of candidates, and the partisan effects obtained by splitting up the total set of candidates in particular, must be prevented.

If a first selection of candidates is made in small nonoverlapping subgroups X_i (i.e., $\bigcup_i X_i = X$ and $X_i \cap X_j = \emptyset \; \forall i, j$) and, in a second phase, the winners of these subgroups are compared, then the final set of winners must be equal to the subset of winners who would have been obtained by comparing the candidates all together.

Figure 5.2(a) is also an example of path independence violation. In fact, if the voting procedure considers successive subgroups of candidates, say, X' and $X - X'$, then some candidates who would be selected as winners when all candidates are considered at once (X) are discarded in the first phase. On the other hand, Figure 5.2(b) is an example of path independence satisfaction. In fact, it can be shown that path independence and the WARP share the strict heritage property, but the latter is stronger than the former, so that a method satisfying WARP also satisfies path independence but not vice versa.

Table 5.1 shows that none of the commonly adopted majoritarian methods satisfies all of the properties described above.

Table 5.1. *The most popular majoritarian methods and the properties they satisfy.*

	First-past-the-post	*Double ballot*	*Approval voting*	*Single transferable vote*	*Amendment procedure*
Anonymity	*	*	*	*	
Condorcet winner					*
Condorcet loser		*		*	*
Pareto principle	*	*		*	
WARP	*		*		
Path independence			*		*

5.2 Proportional Methods

As mentioned in Chapter 2, divisor (or highest average) methods and largest remainder methods are the two groups in which proportional methods are traditionally classified. The difference between these two groups is essentially determined by the procedure followed to assign the seats. However, it can be shown that the properties these two classes satisfy are radically different. A major source of material in this field is the work of Balinski and Young (about 10 articles from 1974 to 1994 and a book in 1982), which contains a detailed discussion of the historical, political, and mathematical aspects of the apportionment problem.

As mentioned in Chapter 2, proportional allocation methods, in the context of electoral systems, have at least five uses:

(1) assignment of seats to cartels according to their votes;

(2) redistribution of seats to lists within cartels according to list votes;

(3) apportionment of seats to districts or regions according to their populations;

(4) in the single transferable vote procedure, in order to distribute the first candidate's excess votes to the second preferences, according to their weight;

(5) redistribution of seats assigned to parties at one level of the district tree when the total number of seats each party is entitled to has already been computed at a higher level of the tree in order to match the number of seats apportioned to the districts at the lower level (this is a *proportional matrix problem*).

The mathematical description of these five problems is the same. In order to give a general formulation, we shall use the language of the classical *ball-box model* of enumerative combinatorics: given S balls and n boxes with given volumes $v_1, \ldots, v_n$, place the balls within the boxes so that the densities (number of balls per unit of volume) in the n boxes are "as constant as possible." While the intuitive meaning of such a sentence is clear, a precise mathematical formulation will be provided in Chapter 6.

Let s_i be the number of balls to be placed in the ith box. Proportional allocation methods attempt to determine $s_1, \ldots, s_n$ so that the densities $\frac{s_1}{v_1}, \ldots, \frac{s_n}{v_n}$ are as close as possible to each other.

Henceforth, we shall use the terms "seats" and "parties" instead of "balls" and "boxes," respectively, to provide the reader with a concrete interpretation; it is understood that any statement can be immediately translated into (1), (2), (3), (4), and (5).

The controversy between divisor and largest remainder methods is reflected in the difficulty of finding a compromise between two very desirable properties: *house monotonicity* and *quota satisfaction*.

House Monotonicity

If the number of available seats (the size of the house) increases from S to $S+1$, house monotonicity guarantees that no party is penalized by receiving fewer seats than those it would have obtained if the total number of seats had not increased. An allocation method satisfies this property if the following condition holds. For any given vote distribution $(v_1, \ldots, v_n)$ and any number S of available seats, if the method assigns s_i seats to party i (where $\sum_{i=1}^{n} s_i = S$), then it must assign $s'_i \geq s_i$ seats to party i when the number of seats increases to $S+1$.

Quota Satisfaction

Quota satisfaction requires that the number of seats each party obtains is as close as possible, either from below or from above, to the exact fraction it deserves, i.e., the exact quota:

$$q_i(\boldsymbol{v}, S) = \frac{v_i S}{\sum_{h=1}^{n} v_h}, \qquad i = 1, 2, \ldots, n.$$

Therefore, a seat assignment method *satisfies quota* (or the Hare property) if the following bounds on the number of seats assigned hold for every party i:

$$q_i - 1 \leq s_i \leq q_i + 1 \quad \forall i.$$

This means that s_i must be equal either to the largest integer smaller than or equal to q_i (denoted by $\lfloor q_i \rfloor$) or to the smallest integer greater than or equal to q_i (denoted by $\lceil q_i \rceil$) as follows:

$$s_i = \lfloor q_i \rfloor \text{ or } s_i = \lceil q_i \rceil.$$

In particular, when q_i is an integer, $s_i = q_i = \lfloor q_i \rfloor = \lceil q_i \rceil$.

This property can be split into two separate requirements, usually known as the Hare minimum and Hare maximum. A method satisfies the *Hare minimum* if

$$q_i - 1 \leq s_i$$

holds for every party i, while it satisfies the *Hare maximum* if

$$s_i \leq q_i + 1$$

holds for every party i.

One of the most discussed issues in the theory of proportional representation is that none of the traditional methods satisfies both house monotonicity and quota satisfaction: divisor methods satisfy the first but not the second, while largest remainders methods satisfy the second but not the first. Many researchers have attempted to design a method that satisfies both properties. In particular, Balinski and Young (1982b) identified the *quota method*, which satisfies both house monotonicity and quota satisfaction.

In the same spirit, more general "quotatone methods" [Still, 1979] satisfying both properties have been essentially derived by adding quota constraints to divisor methods. If, in particular, d'Hondt's method is chosen, then the "quota" method is obtained. However, the quota method did not obtain the expected success and, in spite of its nice theoretical properties, it has been criticized by some authors [Birkhoff, 1976; Carter, 1982]. It was pointed out that quota violations for divisor methods do not occur very often in practice. Furthermore, even if the quota method completely forbids solutions where $|s_i - q_i| > 1$, values close to 1 occur with rather high frequency. In other words, quota violations are eliminated at the expense of a much higher average deviation of s_i from its exact quota q_i.

Other interesting properties have been defined to evaluate and classify the different proportional methods. The following paragraphs highlight the main ideas underlying such properties.

Population Monotonicity

Another monotonicity property that has proved to be quite useful (if not essential in the apportionment problem) relates to the number of seats assigned to a party or state when its weight grows with respect to all others in terms of votes or population. If the votes cast for party i increase while the votes for all other parties remain the same, then party i will expect to get at least the same number of votes it would have obtained if its vote has remained unchanged. Such situations might hardly occur in practice since typically between two different elections the number of votes cast for all parties changes. Therefore, a more appealing definition of population monotonicity is that a party (or state) with a growing weight cannot lose a seat in favor of a party (or state) with a declining weight.

To be more precise, consider two problems, with vote (or population) distribution equal to $(p_1, p_2, \ldots, p_n)$ and S available seats the first problem and $(p'_1, p'_2, \ldots, p'_n)$ and S' the second problem. A method is *population monotone* if for every pair (i, h) such that $p_i \leq p'_i$ and $p_h \geq p'_h$ one of the following conditions necessarily holds:

(1) the number of seats the growing party i receives does not decrease, i.e., $s'_i \geq s_i$;

(2) the number of seats the declining party h receives does not increase, i.e., $s'_h \leq s_h$;

(3) everything (weights and seats) is unchanged for both parties; therefore, $p'_i = p_i$, $p'_h = p_h$, $s_i = s'_i$, and $s_h = s'_h$.

Droop Minimum

Suppose that there are P voters and S seats to be assigned. One would expect that a candidate is assigned a seat only if he or she receives at least $\frac{P}{S}$ votes. Actually, it was pointed out by Henry Droop (1868) that it is enough to get more than $\frac{P}{S+1}$ votes in order to obtain a seat since it is impossible that, given that the total number of votes is P, other S candidates receive $\frac{P}{S+1}$ votes each, assuming that the S seats are obtained by the S most voted candidates. For example, if $P = 100$ and $S = 5$, a candidate who obtains 17 ($> \frac{100}{6}$) votes must win a seat because he or she cannot turn out to be more than fifth in the rank of the most voted candidates. In fact, in order to be ranked sixth, five other candidates must have received at least as many votes, but this is impossible because $17 \times 6 = 102$, while the total number of votes is only 100.

More generally, if the number of votes cast for party i is greater than a multiple of the ratio $\frac{P}{S+1}$, say $a\frac{P}{S+1}$, then we could require that at least a seats be assigned to such party. In fact, the *Droop quota* of a party i is defined as

$$D_i(v, S) = v_i \frac{(S+1)}{P}.$$

If this ratio is not an integer, then if party i gets v_i votes it must obtain at least $\lfloor v_i \frac{(S+1)}{P} \rfloor$ seats. If it is an integer the same party will receive at least $\lfloor v_i \frac{(S+1)}{P} \rfloor - 1$ seats.

In general, an allocation method satisfies the Droop minimum if for every party i

$$s_i \geq \left\lfloor v_i \frac{(S+1)}{P} \right\rfloor,$$

whether the Droop quota is an integer or not.

Consistency

A fundamental property underlying proportionality is *consistency*, which guarantees that any partial assignment is itself proportional. Suppose that there are two parties i and h with p_i and p_h votes, respectively, and that an allocation method, when the number of seats increases from S to $S+1$, gives the additional seat to party i. Suppose now that there is some other distribution of votes where i and h still get p_i and p_h votes each, although the number of votes received by other parties may change. If the number of seats increases from, say, S' to $S'+1$, then the method, to be consistent, should still give the additional seat to party i.

Consistency is a strong property that clearly encourages the use of divisor methods. Typically, divisor methods are consistent since the rule adopted to assign seats is based on the direct comparison between the priority of each party in terms of gaining an additional seat and such priority does not change when a subgroup of parties is considered instead. On the other hand, the Largest Remainders method is not consistent because the priority to receive an extra seat is totally determined by the order in which parties are ranked with respect to their remainders $r_i = v_i \frac{S}{P} - \lfloor v_i \frac{S}{P} \rfloor \, \forall i$. By changing the number of available seats, even if the total vote remains unchanged, the relationships between these fractional remainders can easily change to the point that party priorities are inverted.

All through their work, Balinski and Young clearly advocate consistency as being the essence of proportionality itself and they identify the deep roots of this concept which go back, maybe, to the most ancient fair resource allocation problem in history, provided by the Babylonian Talmud [Young, 1987].

Stability

Suppose an electoral procedure assigns s' seats to a party which has received v' votes and s'' seats to another party with v'' votes. If these two parties merge by sharing a total of $v' + v''$ votes, a *stable* electoral procedure will assign a number s of seats to the "new" party such that

$$s' + s'' - 1 \leq s \leq s' + s'' + 1.$$

In most cases stability is not considered so important because one is interested in understanding which electoral formulas encourage parties to merge or form coalitions rather than to splinter or compete under separate flags. Allocation methods that encourage coalitions (or *superadditive* methods) are those that assure that merging parties never have to give up seats they would have obtained by competing separately ($s \geq s' + s''$), while methods that encourage schisms do the opposite ($s \leq s' + s''$). Notice that this kind of analysis does not take into account the effect a coalition or a schism creates on the vote distribution. In fact, the number of voters willing to vote for two separate parties is usually not the same as the number of voters willing to vote for a coalition between those parties. However, we are only interested in the technical capability of seat allocation methods to modify the established power relationships given the vote distribution.

The Largest Remainders method is stable, as are all divisor methods based on divisor criteria $d(s)$ satisfying the following inequalities for every pair of parties i and h:

$$d(s_i + s_h) \leq d(s_i) + d(s_h) \leq d(s_i + s_h + 1).$$

Table 5.2 indicates which of the above properties are satisfied by the main proportional formulas.

Table 5.2. *Proportional methods and their properties.*

	Largest Remainders	*Divisor methods*	*Quota method*
Anonymity	*	*	*
House monotonicity		*	*
Quota satisfaction	*		*
Population monotonicity		*	
Consistency	*	*	*(2)
Stability	*	*(1)	*

(1) Stability holds for all divisor methods based on a divisor criterion $d(s)$ such that $d(s_i + s_h) \leq d(s_i) + d(s_h) \leq d(s_i + s_h + 1)$ [Balinski and Young, 1982a].

(2) Consistency holds only when restricted to pairs of eligible parties.

Chapter 6

An Integer Optimization Approach to Electoral Formulas

In the previous chapters we have examined several different seat assignment methods. Even though these methods are usually referred to as "formulas" translating votes into seats, they cannot generally be expressed by simple analytical formulas. An electoral formula can be thought of as a *black box* that receives an input (the number of votes each party has obtained and a fixed number of seats S to assign) and it produces an output (a distribution of the seats among the parties). What happens inside the black box is better described by a list of elementary operations to be repeated until all seats have been assigned, rather than by a cumbersome mathematical formula in the usual sense of this term.

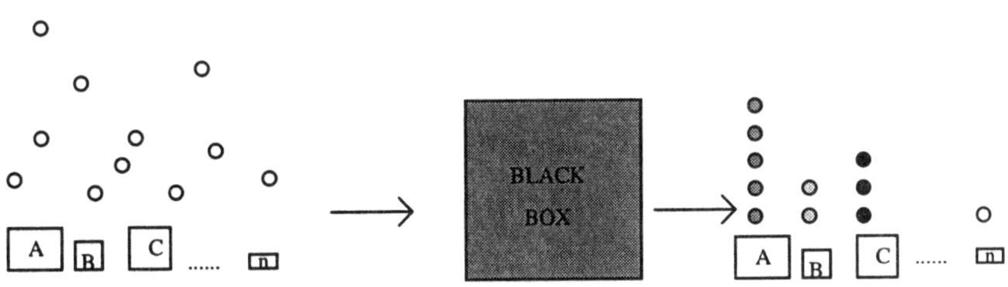

Fig. 6.1. *The vote-to-seat transformation viewed as allocating S balls to n boxes of different volumes.*

In other words, electoral formulas are algorithms and, in our opinion, an algorithmic language is the best to use when defining an electoral formula. This will be shown in the examples provided in the following pages. We have already noticed that the traditional approach to the analysis of electoral systems lacks a unified framework containing both majority and proportional formulas. The algorithmic approach we propose will offer a common language to describe both of these kinds of methods. The language of algorithms is an elegant solution offering a formal description of electoral systems which avoids unnecessary and complicated notation, reduces as much as possible long and confusing sentences, and, therefore, allows a direct comprehension of what happens when votes are transformed into seats.

But this is not the only reason for which an algorithmic approach is highly recommended. In fact, as we will show in detail, electoral formulas are algorithms that solve particular *integer programming problems* with specified objective functions.

6.1 Proportional and Majoritarian Algorithms

Let v_{ij} be the number of votes party i has obtained in district j, n be the total number of parties, S be the number of seats at stake, and s_i be the number of seats assigned to party i. Figures 6.2 to 6.6 fully describe the most common electoral formulas in a procedural language. Notice that for majority formulas we have chosen to refer to a territory divided into S single-member districts, one for each of the available seats. On the other hand, proportional formulas refer to assigning all of the S seats in a unique multimember district. Also notice that in many cases an algorithm that has already been defined is used as a subroutine to define a more complicated electoral formula.

In Chapter 2, divisor methods were described as those selecting the S largest ratios $\frac{v_i}{d(s)}$, $s = 0, 1, \ldots, S-1$; $i = 1, 2, \ldots, n$ and assigning one seat to party i for each selected ratio. The algorithm described in Figure 6.5 does the same thing. This can be justified as follows. Consider nS possibly repeated numbers. Put them one at a time into a *bag* according to the following procedure: initially the bag is empty; at each step, place into the bag the largest number that has still not been selected, breaking ties arbitrarily. After s steps, $s = 1, 2, \ldots, S$, the elements in the bag—repetitions being allowed—are the s largest ones. To complete the argument, recall that in our case, the nS ratios $\frac{v_i}{d(s)}$, $s = 0, 1, \ldots, S-1$; $i = 1, 2, \ldots, n$ are such that

$$\frac{v_i}{d(0)} < \frac{v_i}{d(1)} < \cdots < \frac{v_i}{d(S-1)},$$

since $d(0) > d(1) > \cdots > d(S-1)$. It follows that the largest element outside the bag at a given step is obtained by considering, for each party i, the first unselected ratio in the order given by $\frac{v_i}{d(0)} < \frac{v_i}{d(1)} < \cdots < \frac{v_i}{d(S-1)}$ and, then, by choosing the maximum among these n ratios. In fact, the first unselected ratio for party i at a given step is the largest among all unselected ratios for the same party, and it cannot be followed, in the above order, by any selected ratio relative to party i.

First-past-the-post $(S, n, \boldsymbol{V})$

Input: The total number of seats (S), the number of parties (n), the matrix containing the votes per party and district

$$\boldsymbol{V} = \{v_{ij},\ i = 1, 2, \ldots, n,\ \ j = 1, 2, \ldots, S\}.$$

Step 0: for $i = 1, 2, \ldots, n$: let $s_i = 0$; let $j = 1$;

Step 1: choose index i^* such that $v_{i^*j} = \max\{v_{hj} : 1 \leq h \leq n\}$;
$s_{i^*} = s_{i^*} + 1$;
$j = j + 1$;

Step 2: if $j = S + 1$ then STOP;
otherwise return to step 1.

Fig. 6.2. *An algorithmic procedure for the first-past-the-post method.*

6.2 The Hidden Criterion

One of the major debates in the theory of electoral systems concerns the analysis of the degree of proportionality obtained by different apportionment methods. One usually computes the values of an appropriate index (such as the ones given in section 3.3) on the basis of the electoral outcomes observed in different times and countries. Theoretical studies have also added to the belief that Largest Remainders is the "most proportional of methods," while d'Hondt is the "less proportional." In this chapter we will propose an integer optimization approach to the proportional representation problem, and we examine its consequences on the debate on proportionality. In particular, we will argue that trying to establish a rank for the different electoral formulas in terms of their degree of proportionality can be totally misleading. The fact is that proportional electoral methods and many measures of disproportionality are linked one to the other in a inextricable way: it can be shown that each formula is an algorithm yielding an integer solution to the proportional representation problem and minimizing a suitable (and not necessarily unique) disproportionality index. On one hand, an integer optimization approach guides us to uncover, a posteriori, the objective function that is the "hidden" criterion each formula minimizes; on the other, it allows us to design new electoral formulas corresponding to suitable measures of disproportionality.

Double ballot $(S, n, \boldsymbol{V}, \tau)$

Input: The total number of seats (S), the number of parties (n), the matrix containing the votes per party and district

$$\boldsymbol{V} = \{v_{ij},\ i = 1, 2, \ldots, n,\ \ j = 1, 2, \ldots, S\},$$

a nonnegative threshold (τ).

Step 0: for $i = 1, 2, \ldots, n$: let $s_i = 0$;
for $j = 1, 2, \ldots, S$:
compute the entrance threshold τ_j^* for the second ballot (if the second ballot is open $\tau_j^* = 0$; if it is partially closed $\tau_j^* = \tau$; if it is totally closed $\tau_j^* = v_{[2]j}$, where $v_{[1]j} \geq v_{[2]j} \geq \cdots \geq v_{[n]j}$);
let $j = 1$;

Step 1: if there exists an index h such that $\dfrac{v_{hj}}{\sum_{i=1}^{n} v_{ij}} > 0.50$ then
choose such index, let it be i^*;
let $s_{i^*} = s_{i^*} + 1$;
otherwise

$$\text{let } R_j = \left\{ h : \frac{v_{hj}}{\sum_{i=1}^{n} v_{ij}} \geq \tau_j^* \right\};$$

fix new elections in district j for the candidates in R_j;
let the outcome be described by the $|R_j| \times 1$ matrix $\boldsymbol{V}^j$;
call first-past-the-post $(1, |R_j|, \boldsymbol{V}^j)$;

Step 2: $j = j + 1$;
if $j = S + 1$ then STOP;
otherwise return to step 1.

Fig. 6.3. *An algorithmic procedure for the double-ballot method.*

Many optimality results can be found scattered around in the literature, but they lack a unitary framework.

Consider n parties and S seats. Let $\boldsymbol{v} = (v_1, v_2, \ldots, v_n)$ be the vector of votes each party receives and $\sum_{i=1}^{n} v_i = P$. Then the proportional representation

Largest Remainders $(S, n, \boldsymbol{v})$

Input:	The total number of seats (S), the number of parties (n), the vector of votes $\boldsymbol{v} = (v_1, v_2, \ldots, v_n)$ obtained by the parties.
Step 0:	Let $J = \{1, 2, \ldots, n\}$; for $i = 1, 2, \ldots, n$: compute $q_i = v_i \frac{S}{P}$ and $r_i = q_i - \lfloor q_i \rfloor$; for $i = 1, 2, \ldots, n$: let $s_i = \lfloor q_i \rfloor$;
Step 1:	if $S - \sum_{i=1}^{n} s_i = 0$ then STOP;
Step 2:	otherwise choose index j^* in J such that $r_{j^*} = \max\{r_h : \ r_h = q_h - \lfloor q_h \rfloor, \ h \in J\}$; $s_{j^*} = s_{j^*} + 1$; delete j^* from J; return to Step 1.

Fig. 6.4. *An algorithmic procedure for the Largest Remainders method.*

Divisor method $(S, n, \boldsymbol{v}, d)$

Input:	The total number of seats (S), the number of parties (n), the vector of votes $\boldsymbol{v} = (v_1, v_2, \ldots, v_n)$ obtained by each party, and the divisor criterion $d(s)$ (a monotone increasing function such that $d(s) \in [s, s+1]$).
Step 0:	for $i = 1, 2, \ldots, n$: let $s_i = 0$; for $i = 1, 2, \ldots, n$: compute $\rho_i = \frac{v_i}{d(s_i)}$; let $l = 0$
Step 1:	$l = l + 1$; choose index j^* such that $\rho_{j^*} = \max\{\rho_h : 1 \leq h \leq n\}$; $s_{j^*} = s_{j^*} + 1$; update $\rho_{j^*} = \frac{v_{j^*}}{d(s_{j^*})}$;
Step 2:	if $S - l = 0$ then STOP. otherwise return to 1.

Fig. 6.5. *An algorithmic procedure for divisor methods.*

Mixed system with total detraction $(S, n, \alpha, \boldsymbol{V})$
(α % first-past-the-post /$(1-\alpha)$ % d'Hondt)

Input: The total number of seats (S), the number of parties (n), the dosage α, the matrix containing the votes per party, and district

$$\boldsymbol{V} = \{v_{ij},\ i = 1, 2, \dots, n,\ j = 1, 2, \dots, S\}.$$

Step 0: for $i = 1, 2, \dots, n$: let $s_i = 0$;
let $J = \emptyset$;
let $M = \lfloor \alpha S \rfloor$;

Step 1: Call first-past-the-post $(M, n, \boldsymbol{V})$ and let $\boldsymbol{s}_{(\text{first-past-the-post})}$ be the corresponding output;

Step 2: for $j = 1, 2, \dots, M$:
let $v_{[1]j}$ and $v_{[2]j}$ be the largest and the second largest v_{ij},
$1 \le i \le n$, resp.;

Step 3: [Total detraction rule]
for $i = 1, 2, \dots, n$:
let $J = \{j : v_{ij} = v_{[1]j}\}$;
compute the net votes

$$v_i^* = \sum_{j=1}^{M} v_{ij} - \sum_{j \in J} (v_{[2]j} + 1);$$

Step 4: Call divisor method $(S - M, n, \boldsymbol{v}^*, d(s) = s + 1)$ and let $\boldsymbol{s}_{(\text{d'Hondt})}$ be the corresponding output;

Step 5: [Total seats]
for $i = 1, 2, \dots, n$: let $s_i = s_{i(\text{first-past-the-post})} + s_{i(\text{d'Hondt})}$.

Fig. 6.6. *An algorithmic procedure for a mixed system with total detraction.*

problem consists of determining $\boldsymbol{s} = (s_1, s_2, \dots, s_n)$ such that

$$\begin{aligned} &\sum_{i=1}^{n} s_i = S, && \\ &s_i \ge 0, \text{integer}, && \forall i = 1, 2, \dots, n, \\ &s_i \approx v_i && \forall i = 1, 2, \dots, n, \end{aligned} \tag{P1}$$

where the symbol "$\approx$" means "the most proportional possible." In order for the number of seats to be exactly proportional to the number of votes, party i must obtain exactly $s_i = v_i \frac{S}{P}$ seats, $i = 1, 2, \ldots, n$. Unfortunately, this amount (called the *exact quota*) is generally fractional and some rounding must take place. Therefore, there is no unique solution to problem (P1).

A seat allocation that is exactly proportional is perfectly fair to all parties. Fairness or neutrality implies that all parties must be treated in the same way and no one be favored in terms of seats if this can damage others. But the integer constraint on the number of seats each party can obtain will generally cause a *distortion* or *inequality* in the allocation. The problem of determining an exactly proportional seat allocation (P1) is the same as trying to find an efficient procedure to minimize the inequality in representation between single parties. To compare parties in terms of unequal representation, it is obviously necessary to define a precise index of inequality, which measures how much a given seat apportionment is "less unequal" than another. For the most common proportional apportionment formulas such a measure of unfairness is easy to find and may be viewed as a precise mathematical interpretation of the representation concept underlying the single methods. The solutions to problem (P1) given by the different apportionment formulas are in other terms optimal solutions to the following optimization problem:

$$\begin{aligned} \min \quad & \varphi(\boldsymbol{s}; \boldsymbol{v}), \\ & \sum_{i=1}^{n} s_i = S, \\ & s_i \geq 0, \text{ integer}, \; i = 1, 2, \ldots, n, \end{aligned} \tag{P2}$$

where the cost function $\varphi(\boldsymbol{s}; \boldsymbol{v})$ is a *measure of unfairness* (or a disproportionality index) specific to each method, which takes nonnegative values and is usually such that $\varphi(\boldsymbol{s}; \boldsymbol{v}) = 0$ if and only if $s_i = v_i \frac{S}{P}$, $i = 1, 2, \ldots, n$ (in some rare cases the minimum is positive, although the above solution is still optimal).

There are quite a few classes of interesting unfairness measures, most of which are based on the L_ρ-norm of some absolute or relative deviation from the ideal apportionment. Table 6.1 presents some typical objective functions $\varphi(\boldsymbol{s}; \boldsymbol{v})$ and the electoral methods that constitute an optimization algorithm for the corresponding seat allocation problem formulated in (P2).

Notice that function (5) takes only nonnegative values, since

$$\min \left\{ \sum_{i=1}^{n} \frac{s_i^2}{v_i} : \sum_{i=1}^{n} s_i = S \right\} = \frac{S^2}{P},$$

as shown by a standard Lagrange multiplier argument. Interchanging s_i and v_i, for each i, one concludes that function (4) is nonnegative, too.

Largest Remainders minimizes functions in class (1) for all $\rho \geq 1$ [Birkhoff, 1976]. The Sainte-Laguë method minimizes (2) for $\rho = 2$ as recalled by many authors [Balinski and Young, 1982a; Ibaraki and Katoh, 1988] and, apparently, this property was known to Sainte-Laguë himself, while Equal Proportions minimizes (3) for $\rho = 2$. The bottleneck functions (8) and (10) are optimized by

Table 6.1. *Disproportionality measures and corresponding allocation methods.*

	$\varphi(\boldsymbol{s};\boldsymbol{v})$	***Seat allocation method***
1	$\sum_{i=1}^{n} \lvert s_i - q_i \rvert^\rho$	Largest Remainders for all $\rho \geq 1$
2	$\sum_{i=1}^{n} v_i \left\lvert \frac{s_i}{v_i} - \frac{S}{P} \right\rvert^\rho$	Largest Remainders for $\rho = 1$, Sainte-Laguë for $\rho = 2$
3	$\sum_{i=1}^{n} s_i \left\lvert \frac{v_i}{s_i} - \frac{P}{S} \right\rvert^\rho$	Equal Proportions for $\rho = 2$
4	$\sum_{i=1}^{n} v_i \left(\frac{v_i}{s_i} - \frac{P}{S} \right)$	Equal Proportions
5	$\sum_{i=1}^{n} s_i \left(\frac{s_i}{v_i} - \frac{S}{P} \right)$	Sainte-Laguë
6	$\sum_{i=1}^{n} \left\lvert \frac{s_i}{S} - \frac{v_i}{P} \right\rvert^\rho$	Largest Remainders for all $\rho \geq 1$
7	$\max_{i=1,2,\ldots,n} \left\lvert \frac{s_i}{S} - \frac{v_i}{P} \right\rvert$	Largest Remainders
8	$\max_{i=1,2,\ldots,n} \frac{s_i}{v_i}$	d'Hondt
9	$\sum_{i=1}^{n} \left(\frac{s_i}{v_i} - \frac{S}{P} \right)^+$, where $z^+ = \max(z, 0)$	d'Hondt
10	$\max_{i=1,2,\ldots,n} \frac{v_i}{s_i}$	Smallest Divisors
11	$\sum_{i=1}^{n} \frac{(s_i - q_i)^2}{q_i}$	Sainte-Laguë
12	$\sum_{i=1}^{n} \frac{(s_i - q_i)^2}{s_i}$	Equal Proportions

the d'Hondt method and Smallest Divisors, respectively [Balinski and Young, 1982a; Ibaraki and Katoh, 1988]. We can also notice that functions in class (6), including function (7) where $\rho \to \infty$, are multiples of those in class (1) and therefore are still minimized by Largest Remainders for all $\rho \geq 1$. The same is true for (2) when $\rho = 1$. The remaining results appear to be new.

6.3 Greedy Algorithms for Proportional Allocation

The proportional representation problem (P2) may be viewed as a discrete resource allocation problem. Given a total amount of resource S (in our case the *seats*) we want to allocate it to n individuals or activities (in our case the *parties*) so that the value of some objective function is as small as possible, considering that the amount of resource s_i assigned to i may take only non-negative integer values (since no party can receive a fraction of seats). The general resource allocation problem is computationally hard to solve, except for some well-behaved classes of objective functions. Consider the following discrete resource allocation problem:

$$\begin{aligned} \min \quad & \varphi(s_1, s_2, \ldots, s_n), \\ & \sum_{i=1}^{n} s_i = S, \\ & s_i \geq 0, \quad \text{integer}, \quad i = 1, 2, \ldots, n. \end{aligned} \tag{6.1}$$

The function $\varphi(s_1, s_2, \ldots, s_n)$ is *separable* if it can be written as the sum of n functions of a single argument:

$$\varphi(s_1, s_2, \ldots, s_n) = \varphi_1(s_1) + \varphi_2(s_2) + \cdots + \varphi_n(s_n).$$

Let $\boldsymbol{e}_i$ be the n-component vector with a unit in position i and zeros elsewhere. The *discrete derivative* in direction i of a function $\varphi(\boldsymbol{s})$, $\boldsymbol{s} \in \Re^n$, is the difference between the value of function φ in the point $\boldsymbol{s} + \boldsymbol{e}_i$ and its value in $\boldsymbol{s}$,

$$\varphi_i(\boldsymbol{s} + \boldsymbol{e}_i) - \varphi(\boldsymbol{s}).$$

Clearly, if $\varphi(\boldsymbol{s})$ is separable, its discrete derivative in direction i is a function of a single argument for every $i = 1, 2, \ldots, n$ and, in particular, $\varphi(\boldsymbol{s}+\boldsymbol{e}_i) - \varphi(\boldsymbol{s}) = \varphi_i(s_i + 1) - \varphi_i(s_i)$. Therefore, function $\varphi(\boldsymbol{s})$ can be written as the sum of a constant and a function of its discrete derivatives as follows:

$$\varphi(\boldsymbol{s}) = \sum_{i=1}^{n} \varphi_i(s_i) = \sum_{i=1}^{n} \left[\varphi_i(0) + h_i(s_i)\right],$$

where

$$h_i(s_i) = \begin{cases} 0 & \text{if } s_i = 0, \\ \rho_i(0) + \rho_i(1) + \cdots + \rho_i(s_i - 1) & \text{if } s_i \geq 1. \end{cases}$$

Henceforth, let $K = \sum_{i=1}^n \varphi_i(0)$ and $h(\boldsymbol{s}) = \sum_{i=1}^n h_i(s_i)$, so that $\varphi(\boldsymbol{s}) = K + h(\boldsymbol{s})$. Now, assume that each discrete derivative ρ_i is nondecreasing:

$$\rho_i(y) \geq \rho_i(y-1), \qquad y = 1, 2, \ldots, S-1.$$

Consider all the nS values $\rho_i(y)$, $i = 1, 2, \ldots, n$; $y = 0, 1, \ldots, S-1$ (some of which may coincide) and an arbitrary nondecreasing ordering of such values with the property that, when $\rho_i(y) = \rho_i(y-1)$, then $\rho_i(y-1)$ precedes $\rho_i(y)$ in this ordering. Consider next the *bag* (or *multiset*) L_S containing the first S values $\rho_i(y)$—with possible repetitions—in the chosen ordering. Notice that, although the ordering might not be unique because of possible ties, the bag L_S is indeed unique if one looks at its elements and their multiplicities. It follows that both the sum l_S and the maximum λ_S of the S elements of L_S are well defined. Clearly, if more than one element in the set of all discrete derivatives has the value λ_S, it could be that not all of them are included in L_S, since $|L_S| = S$.

The following results hold for any separable function $\varphi(\boldsymbol{s})$ with nondecreasing discrete derivatives ($\rho_i(0) \leq \rho_i(1) \leq \cdots \leq \rho_i(S-1)\ \forall i$), providing a simple method to solve the integer resource allocation problem with such an objective function.

LEMMA 6.1. *If $\varphi(\boldsymbol{s})$ is a separable function and its discrete derivative $\rho_i(\boldsymbol{s})$ is nondecreasing for every $i = 1, 2, \ldots, n$, then $\varphi(\boldsymbol{s}) \geq K + l_S$ for every allocation $\boldsymbol{s}$ satisfying the cardinality constraint $\sum_{i=1}^n s_i = S$.*

Proof. Since $\varphi(\boldsymbol{s}) = K + h(\boldsymbol{s})$, all we have to show is that $h(\boldsymbol{s}) \geq l_S\ \forall \boldsymbol{s}$, where $\sum_{i=1}^n s_i = S$. By construction, L_S contains the S smallest values (possibly repeated) observed in the set $\{\rho_i(y) : 0 \leq y \leq S-1,\ 1 \leq i \leq n\}$; therefore, $l_S \leq h(\boldsymbol{s})$ whatever the particular $\boldsymbol{s}$ under consideration. □

LEMMA 6.2. *If there exists an allocation $\boldsymbol{s}^*$ such that $h(\boldsymbol{s}^*) = l_S$, then $\boldsymbol{s}^*$ is an optimal solution to the discrete resource allocation problem with $\varphi(\boldsymbol{s})$ as objective.*

Proof. This is obvious since minimizing $\varphi(\boldsymbol{s})$ is equivalent to minimizing $h(\boldsymbol{s})$. □

LEMMA 6.3. *If $\boldsymbol{s}$ is an allocation such that $h(\boldsymbol{s}) > l_S$, then there exist two parties i and j, with $s_i \geq 1$, such that*

$$\rho_j(s_j) < \rho_i(s_i - 1).$$

Proof. Consider the subbag[5] of L_S given by $L_S^- \equiv \{\rho_h(y) \in L_S : \rho_h(y) < \lambda_S\}$. Notice that L_S is formed by the $p = |L_S^-|$ elements of L_S^- and by $(|S| - p) \geq 1$

[5]A bag B' is a *subbag* of the bag B if each element of B' is also an element of B and its multiplicity in B' does not exceed its multiplicity in B.

occurrences of λ_S. Now let s be an allocation such that $h(s) > l_S$. Since some party must get at least one seat, the bag

$$H(s) = \{\rho_h(y) : s_h \geq 1; y = 0, 1, \dots, s_h - 1\}$$

is nonempty. One has $|H(s)| = \sum_{h=1}^{n} s_h = S$ and $h(s)$ is precisely the sum of all elements of $H(s)$, counted with their multiplicities. Let μ be the maximum element of $H(s)$. Since the derivatives are nondecreasing, $\mu = \rho_i(s_i - 1)$ for some party i such that $s_i \geq 1$. Moreover, $\mu \geq \lambda_S$, otherwise $H(s)$ would contain at most $p < S$ elements. We claim that there must be a party j and an integer, $y \geq 0$ such that $\rho_j(y) \in L_S - H(s)$ and $\rho_j(y) < \mu$. There are two cases.

Case 1. $H(s) \supseteq L_S^-$. In this case one must have $\mu > \lambda_S$; otherwise, both L_S and $H(s)$ would consist of the p elements of L_S^- and by $S - p$ occurrences of λ_S, implying $h(s) = l_S$. But then, since at most $S - 1$ elements of $H(s)$ belong to L_S and $|L_S| = |H(s)|$, there must be some $\rho_j(y) \in L_S - H(s)$, and one has $\rho_j(y) \leq \lambda_S < \mu$.

Case 2. *There exists a* $\rho_j(y) \in L_S^- - H(s)$. In this case $\rho_j(y) < \lambda_S \leq \mu$. Thus the claim is proved in any case. If $s_j = 0$, then $\rho_j(s_j) \leq \rho_j(y) < \rho_i(s_i - 1)$. If $s_j \geq 1$, then one cannot have $s_j - 1 \geq y$, else $\rho_i(y)$ would belong to $H(s)$. Hence, $s_j \leq y$, implying that $\rho_j(s_j) \leq \rho_j(y) < \rho_i(s_i - 1)$. Thus the lemma has been proved. □

A procedure works in a *greedy* fashion if it is only concerned with *marginal allocations*; that is, it chooses the allocation that minimizes (or maximizes) the objective function at each step. More precisely, the greedy algorithm is defined as follows. Starting from the allocation $s = (0, 0, \dots, 0)$, at each step the algorithm assigns one more seat to a party i such that, for the current allocation $(s_1, s_2, \dots, s_n)$, one has $\rho_i(s_i) = \min\{\rho_h(s_h) : 1 \leq h \leq n\}$. The algorithm stops when all the S seats have been assigned. This is equivalent to assigning a seat to the S smallest elements of the bag $\{\rho_i(y) : y = 0, 1, \dots, S - 1;\ i = 1, 2, \dots, n\}$.

THEOREM 6.1. *If $\varphi(s)$ is a separable function and its discrete derivative $\rho_i(s)$ is nondecreasing for every $i = 1, 2, \dots, n$, then the greedy procedure provides an optimal solution for the discrete resource allocation problem with $\varphi(s)$ as the objective.*

Proof. Let s^* be the solution obtained by the greedy procedure. By Lemma 6.1, $h(s^*) \geq l_S$. Suppose that strict inequality holds; that is,

$$h(s^*) > l_S. \tag{6.2}$$

By Lemma 6.3, this implies that there exist two parties i and j, with $s_i \geq 1$, such that $\rho_j(s_j) < \rho_i(s_i - 1)$. However, in this case, s^* would not be the solution given by the greedy procedure, because a greedy algorithm would have preferred to assign an extra seat to party j instead of party h, giving j a total of $s_j^* + 1$ seats instead of s_j^* and h a total of $s_h^* - 1$ instead of s_h^*. □

Our proof follows an approach similar to the one in Ibaraki and Katoh (1988) but it relies, rather than on Lagrangian duality, on a more direct combinatorial argument. In particular, Theorem 6.1 holds when all functions φ_i are convex,[6] since then their discrete derivatives are nondecreasing. In this case, the theorem was first established in Gross (1956). It is worth mentioning that a generalization of Theorem 6.1 to the minimization of a *quasi-separable pseudo-convex* objective function over an integral polymatroid has been obtained by Camerini, Conforti, and Naddef (1989).

The greedy procedure is an optimization algorithm for a wide class of objective functions listed in Table 6.1. The attentive reader will have surely not missed the strong resemblance between greedy algorithms and divisor methods (see Figure 6.5). As a matter of fact, both are iterative procedures that allocate one seat at a time, so as to achieve at each step the maximum possible "profit" of a party. In greedy algorithms the profit is just (the absolute value of) the decrease of the inequality measure under consideration. As for divisor methods, the profit is defined by the ratio $\frac{v_i}{d(s_i)}$. Recall that the divisor criterion always satisfies the inequalities $x \leq d(x) \leq x + 1$. The two extremes, $d(x) = x + 1$ and $d(x) = x$, correspond to d'Hondt's and to Adams's method, respectively. D'Hondt's method maximizes $\frac{v_i}{s_i+1}$, the average number of votes per seat *after* the allocation of the current seat, while Adams's method maximizes $\frac{v_i}{s_i}$, the average number of votes per seat *before* the allocation of the seat, instead. Any other divisor method is such that $\frac{v_i}{s_i+1} \leq \frac{v_i}{d(s_i)} \leq \frac{v_i}{s_i}$. Imagine that the number of seats of party i changes from s_i to $s_i + 1$ continuously. Somewhere during this transformation the ratio votes to seats is equal to $\frac{v_i}{d(s_i)}$. To summarize, the profit in any divisor method can always be interpreted as an average number of votes per seat. In conclusion, if one more generally defines as "greedy" an algorithm that adds elements to a set one at a time so as to achieve at each step the largest profit (for a suitable profit criterion), then divisor methods may be indeed regarded as greedy.

The greedy allocation method is a unifying factor in the proof of optimality results for proportional allocation methods. In Table 6.2, the discrete derivative of the relevant fairness measures is considered and the corresponding divisor criteria are identified.

However, divisor methods follow a greedy procedure which does not guarantee quota satisfaction in the final seat assignment. Nevertheless, for some of the objective functions listed in Table 6.1 quota satisfaction is a necessary condition in order to attain the minimum possible value under the allocation constraints (functions with this property will be called *quota property functions*). For these functions, given any solution that does not respect the lower and upper bounds, one can always find some better solution in which $\lfloor q_i \rfloor \leq s_i \leq \lceil q_i \rceil$.

[6] A subset C of $\Re^n$ is called *convex* if, whenever it contains two points x and y, it contains $\alpha\mathsf{x} + (1-\alpha)\mathsf{y}$ for all $0 \leq \alpha \leq 1$. A real-valued function f defined in the convex set C is *convex* if, for all $\mathsf{x}, \mathsf{y} \in C$ and $0 \leq \alpha \leq 1$, one has $f(\alpha\mathsf{x} + (1-\alpha)\mathsf{y}) \leq \alpha f(\mathsf{x}) + (1-\alpha) f(\mathsf{y})$.

Table 6.2. *Objective functions and corresponding divisor methods.*

$\varphi(s;v)$	$\rho_i(s_i)$	***Divisor criterion*** $d(s)$
$\sum_{i=1}^{n} v_i\left(\frac{s_i}{v_i}-\frac{S}{P}\right)^2$	$\frac{2s_i+1}{v_i}-\frac{2S}{P}$	$s+\frac{1}{2}$ Sainte-Laguë
$\sum_{i=1}^{n} s_i\left(\frac{v_i}{s_i}-\frac{P}{S}\right)^2$	$\frac{P^2}{S^2}-\frac{v_i^2}{s_i(s_i+1)}$	$\sqrt{s(s+1)}$ Equal Proportions
$\sum_{i=1}^{n} v_i\left(\frac{v_i}{s_i}-\frac{P}{S}\right)$	$-\frac{v_i^2}{s_i(s_i+1)}$	$\sqrt{s(s+1)}$ Equal Proportions
$\sum_{i=1}^{n} s_i\left(\frac{s_i}{v_i}-\frac{S}{P}\right)$	$\frac{2s_i+1}{v_i}-\frac{S}{P}$	$s+\frac{1}{2}$ Sainte-Laguë
$\sum_{i=1}^{n}\left(\frac{s_i}{v_i}-\frac{S}{P}\right)^+$	$\begin{cases}\frac{1}{v_i} & \text{if } \frac{s_i}{v_i}-\frac{S}{P}\geq 0\\ 0 & \text{if } \frac{s_i+1}{v_i}-\frac{S}{P}\leq 0\\ \frac{s_i+1}{v_i}-\frac{S}{P} & \text{otherwise.}\end{cases}$	$s+1$ d'Hondt
$\sum_{i=1}^{n}\frac{(s_i-q_i)^2}{q_i}$	$\frac{P}{S}\frac{2s_i+1}{v_i}-2$	$s+\frac{1}{2}$ Sainte-Laguë
$\sum_{i=1}^{n}\frac{(s_i-q_i)^2}{s_i}$	$1-\frac{S^2}{P^2}\frac{v_i^2}{s_i(s_i+1)}$	$\sqrt{s(s+1)}$ Equal Proportions

Consider the following discrete optimization problem:

$$\begin{aligned}&\min\ \psi(s_1,s_2,\ldots,s_n),\\ &\sum_{i=1}^{n} s_i = S, \qquad\qquad (6.3)\\ &s_i\geq 0,\quad \text{integer},\quad i=1,2,\ldots,n,\end{aligned}$$

where $\psi(s_1,s_2,\ldots,s_n)$ is a quota property function. Then there is an optimal solution $s\in\mathfrak{R}^n$ to problem (6.3) which can be written as

$$s_i=\lfloor q_i\rfloor+z_i,\quad i=1,2,\ldots,n,$$

where z_i takes the values 0 or 1.

Suppose, furthermore, that the function ψ is separable. Then one has

$$\psi_i(s_i) = \psi_i(\lfloor q_i \rfloor + z_i) = (1-z_i)\psi_i(\lfloor q_i \rfloor) + z_i\psi_i(\lfloor q_i \rfloor + 1) = c_i z_i + b_i, \quad i = 1, 2, \ldots, n,$$

where $c_i = \psi_i(\lfloor q_i \rfloor + 1) - \psi_i(\lfloor q_i \rfloor)$ and $b_i = \psi_i(\lfloor q_i \rfloor)$. Therefore, the above optimization problem is equivalent to

$$\begin{aligned} \min\ & c_1 z_1 + c_2 z_2 + \cdots + c_n z_n, \\ & z_1 + z_2 + \cdots + z_n = R \equiv S - \sum_{i=1}^{n} \lfloor q_i \rfloor, \\ & z_i \in \{0, 1\}, \qquad i = 1, 2, \ldots, n. \end{aligned}$$

The latter problem is a *cardinality-constrained linear binary knapsack problem*, which can be easily solved by the following *greedy* rule: set to 1 all the variables z_i corresponding to the smallest R coefficients c_i; set to 0 all remaining variables z_i. For example, consider the function (1) with $\rho = 1$. This function clearly satisfies quota. In fact, assume that $\boldsymbol{s} = (s_1, s_2, \ldots, s_n)$ is a seat allocation that does not satisfy quota. Then one of the two following cases must occur:

(1) there is an i such that $s_i < \lfloor q_i \rfloor$;

(2) there is an i such that $s_i > \lceil q_i \rceil$.

In the first case there must also exist a j such that $s_j \geq \lceil q_j \rceil$, for otherwise one would have

$$S = \sum_{t=1}^{n} s_t < \sum_{t=1}^{n} \lfloor q_t \rfloor \leq S,$$

which is clearly a contradiction.

Set

$$s'_h = \begin{cases} s_h + 1, & h = i, \\ s_h - 1, & h = j, \\ s_h, & h \neq i, j. \end{cases}$$

Then one can check that $\psi(\boldsymbol{s}'; \boldsymbol{v}) \leq \psi(\boldsymbol{s}; \boldsymbol{v})$. Iterating this procedure, one eventually finds a solution $\boldsymbol{s}^*$, which is better than $\boldsymbol{s}$ and satisfies quota. A similar argument holds true in the second case.

Moreover, ψ is separable. The binary knapsack problem in this case is equivalent to

$$\begin{aligned} \max\ & r_1 z_1 + r_2 z_2 + \cdots + r_n z_n, \\ & z_1 + z_2 + \cdots + z_n = R, \\ & z_i \in \{0, 1\}, \quad i = 1, 2, \ldots, n, \end{aligned}$$

where r_i is the remainder $q_i - \lfloor q_i \rfloor$. This follows from the identities $c_i = (\lfloor q_i \rfloor + 1 - q_i) - (q_i - \lfloor q_i \rfloor) = 1 - 2r_i$, $i = 1, 2, \ldots, n$. Notice that in this case the greedy rule yields precisely the largest remainders method.

It can also be shown that all divisor methods yield optimal solutions with respect to some *bottleneck function* as stated in the following theorem.

In what follows, we shall make the assumption that the divisor criterion $d(s)$ takes only strictly positive values. In expressions such as $\min_i \frac{v_i}{d(s_i-1)}$, it is understood that the minimum is taken over all parties i such that $s_i \geq 1$. Even in divisor methods (such as Adams's) in which $d(0) = 0$, the theorem below and its proof remain valid, provided that one makes the convention that $\frac{v_i}{0} = +\infty$.

THEOREM 6.2. *Any divisor method with a positive increasing divisor criterion $d(s)$ provides an optimal solution to the proportional representation problem with the following objective:*

$$\max_s \min_i \frac{v_i}{d(s_{i-1})}.$$

Proof. Let $\boldsymbol{s}^*$ be the solution obtained by applying a divisor method with divisor criterion $d(s)$ and let

$$\frac{v_l}{d(s_l^* - 1)} = \min_i \frac{v_i}{d(s_i^* - 1)}.$$

Notice that, whatever the divisor method, the last ratio taken into consideration when allocating the S seats is precisely $\frac{v_i}{d(s_i^*-1)}$ for every i, if $(s_1^*, s_2^*, \ldots, s_n^*)$ is the final allocation. Suppose that there is some other allocation $\boldsymbol{s} \neq \boldsymbol{s}^*$ still satisfying the constraints of problem (P2) and, in particular,

$$\sum_{i=1}^{n} s_i = \sum_{i=1}^{n} s_i^* = S, \tag{6.4}$$

which provides a greater value for the objective function. This means that there exists a party h such that

$$\frac{v_h}{d(s_h - 1)} = \min_i \frac{v_i}{d(s_i - 1)} > \min_i \frac{v_i}{d(s_i^* - 1)} = \frac{v_l}{d(s_l^* - 1)}. \tag{6.5}$$

Since the divisor criterion is a monotone increasing function, it must be that $s_l < s_l^*$. In fact, if it were $s_l \geq s_l^*$ then $d(s_l - 1) \geq d(s_l^* - 1)$ and therefore

$$\frac{v_h}{d(s_h - 1)} > \frac{v_l}{d(s_l^* - 1)} \geq \frac{v_l}{d(s_l - 1)},$$

which is clearly in contradiction with the fact that $\frac{v_h}{d(s_h-1)} = \min_i \frac{v_i}{d(s_i-1)}$ as stated in (6.5). Moreover, if $s_l < s_l^*$, since the constraint (6.4) must hold for $\boldsymbol{s}$, there must exist a party, say, party j, such that $s_j > s_j^*$ and since all variables are integers this means $s_j \geq s_j^* + 1$.

At this point the following lemma by Balinski and Young (1982a) is used.

LEMMA. *A seat allocation method is a divisor method if and only if it provides an allocation* $\mathbf{t}$ *such that*

$$\min_i \frac{v_i}{d(t_i - 1)} \geq \max_i \frac{v_i}{d(t_i)}. \tag{6.6}$$

Therefore, applying (6.6) and recalling that $s_j \geq s_j^* + 1$ (i.e., $s_j - 1 \geq s_j^*$) we get

$$\frac{v_l}{d(s_l^* - 1)} = \min_i \frac{v_i}{d(s_i^* - 1)} \geq \max_i \frac{v_i}{d(s_i^*)} \geq \frac{v_j}{d(s_j^*)} \geq \frac{v_j}{d(s_j - 1)} \tag{6.7}$$
$$\geq \min_i \frac{v_i}{d(s_i - 1)} = \frac{v_h}{d(s_h - 1)} > \frac{v_l}{d(s_l^* - 1)}.$$

We must therefore conclude that (6.5) cannot hold. □

6.4 Consequences of an Optimization Approach to Proportional Representation

In the previous paragraphs we have basically shown that proportional electoral formulas are algorithms and that they minimize specific and not necessarily unique objective functions. Each objective function can be considered as a hidden criterion underlying the corresponding proportional formula. Besides the fact that it offers an algorithmic framework for many proportional methods via greedy allocation procedures, such an approach has at least four important consequences on the evaluation and on the design of electoral systems, which can be briefly stated as follows:

(1) it provides a nonambiguous interpretation of the difference between proportional methods, which can be easily translated into political terms to understand the different concept of representation each method follows and to explain trends observed on their behavior;

(2) it proves that the traditional analysis of disproportionality, based on comparing the values of an appropriate disproportionality index to establish a rank of the different methods, is usually biased in favor of the Largest Remainders method, thus leading to the general belief that such a method is the most proportional;

(3) it allows us to design new electoral methods by choosing some other unfairness or disproportionality criteria which may turn out to be just as good as the well-known ones;

(4) it may be used to compare proportional methods on the basis of their complexity (as will be further discussed in section 6.5).

First, let us show how the objective functions listed in Table 6.1 offer a useful insight on the different meaning each electoral formula assigns to the concept of representation. The reader is warned that, in the literature, the same allocation method is referred to with various names depending on the specific national context. Many proportional formulas have been proposed, designed, and reinvented in history by different mathematicians and politicians on the basis of different grounds, creating some confusion in the general terminology. In Table 6.3, we display some of the different names used equivalently to denote the same method.

Table 6.3. *The many names used for traditional proportional methods.*

1	Greatest Divisors	d'Hondt	Jefferson		Highest Average
2	Smallest Divisors		Adams		
3	Equal Proportions	Huntington	Hill	Geometric Mean	
4	Harmonic Mean		Dean		
5	Major Fractions	Sainte-Laguë	Webster	Arithmetic Mean	Odd Numbers
6	Largest Remainders	Vinton	Hamilton	Hare	

The Method of Sainte-Laguë

The method proposed in 1910 by Sainte-Laguë in a French review (and actually suggested several years before by Webster) was designed with the intention of guaranteeing each elector the same power or portion of representation. When $\rho = 2$, objective function (2) measures the disproportionality of the seat allocation in terms of the power of each single voter. Fairness between voters means that all of the P voters must have the same power or influence on the electoral result, so that if the total number of representatives is S, each voter must have power on $\frac{S}{P}$ representatives. On the other hand, the v_i voters that cast their vote for party i actually have power on $\frac{s_i}{v_i}$ representatives, since s_i is the number of seats allocated to party i. Therefore, the differences between $\frac{s_i}{v_i}$ and $\frac{S}{P}$ determine each voter's *error* or *deviation* from the ideal allocation. It is natural to believe that if such deviations are caused by a neutral method, they must be allowed only if they are random and not systematic. The Sainte-Laguë divisor method is "fair" because it minimizes these deviations under the hypothesis of a Gaussian or normal distribution of the errors.

Notice that minimizing function (2) with $\rho = 2$ is equivalent to minimizing function (11)

$$\sum_{i=1}^{n} \frac{(s_i - q_i)^2}{q_i},$$

where q_i is party i's exact quota, since they only differ by the positive multiple $(\frac{P}{S})$. This function represents the sum of the squared deviations between the actual number of seats assigned to each party and the ideal number of seats, divided by this ideal number. Relative errors of this kind are a classic in evaluating approximate solutions. On the other hand, when $s_i + \cdots + s_n = S$, function (5) differs from function (2) with $p = 2$ by an additive constant, and thus is minimized by the Saint-Laguë method as well.

The Method of Equal Proportions

The method of Equal Proportions has been used for over 50 years in the United States (where it is called Hill's method) to assign representatives to the different states on the basis of their population. It has not received the same success in Europe where the problem usually is that of assigning seats to parties on the basis of their votes. The reason is that it automatically assigns each party at least one seat whatever its number of votes and thus tends to produce a very high fractionalization of the seats in the legislature. Bringing the first divisor from 0 to a small positive value would avoid this inconvenience and make this method very appropriate for the European context, too.

Three distinct objective functions have been listed for this method in Table 6.1. For all of them, notice that the quantity $\frac{v_i}{s_i}$ is the *price*, in terms of votes, at which party i has bought one seat. If all parties are treated in the same way, the cost per seat should be the same for all parties. In fact, this cost should be equal to $\frac{P}{S}$.

Equal Proportions minimizes the relative error between the actual seat allocation and the ideal one, normalized with respect to the number of seats actually allocated to each party (function (12)).

Furthermore, function (4) according to Equal Proportions minimizes a weighted sum of the differences between the price per seat each party pays and the actual cost of a seat, where an extra-cost (positive deviations) is rated more undesirable the larger the party is, since the weight assigned to each deviation is equal to $v_i, \quad i = 1, 2, \ldots, n$. Notice that, since $s_1 + s_2 + \cdots + s_n = S$, function (3) with $\rho = 2$ differs from function (4) by an additive constant, and from (12) by a positive constant factor.

The Largest Remainders Method

The fairness criterion embodied in the Largest Remainders method is probably the most widely accepted, and in any case it generates all traditional disproportionality indicators. Function (1) is the L_ρ-norm of the simple deviations between the number of seats that are actually assigned and the corresponding exact quotas. Function (6) is basically the same but this time deviations between vote shares and seat shares are directly considered. In particular, Largest Remainders will always provide a seat assignment that minimizes the maximum deviation between vote and seat share (i.e., the L_∞-norm showing up in function (7)).

Finally, function (2) with $\rho = 1$ is the weighted sum of the absolute distances between party i's current degree of representation $(\frac{s_i}{v_i})$ and a perfectly proportional representation, where a violation from the ideal $\frac{S}{P}$ is considered more unfair for a party with a larger number of votes, since the weight assigned is $v_i, \quad i = 1, 2, \ldots, n$.

The d'Hondt Method

The d'Hondt method is one of the most commonly used methods in Europe. It has been sharply criticized for being the "less proportional method." However, the integer optimization approach we suggest adopting shows that d'Hondt embodies a very important proportionality criterion. The seat allocation provided by the d'Hondt method maximizes the bottleneck function (8), which concerns the minimum price paid by a party in order to obtain one seat ($\frac{v_i}{s_i}$). If a seat allocation gives the maximum value for this sort of function, it means that the difference between the cost per seat paid by the different parties is smoothed as much as possible. It seems desirable that these costs be eventually all equal in a fair seat assignment.

Moreover, such optimality property provides a direct explanation to the fact that d'Hondt tends to advantage bigger parties and to be severe with small ones. In fact, forcing the minimum price per seat to reach its highest possible value will automatically cut out the smaller parties that will not be able to pay it.

The d'Hondt method also minimizes the Schultz index (function (9)), used in economics to measure income inequality. In this case, the function measures the overall amount of overrepresentation the seat allocation has created. Therefore, although some party will have to be overrepresented since the exact quota cannot be assigned, the idea of proportionality embodied in this method guarantees that, among all allocations, the one yielding the minimum possible total overrepresentation is chosen.

The Method of Smallest Divisors

The rationale underlying the method of Smallest Divisors is complementary to the one embodied in d'Hondt. In fact, this method will analogously attempt to smooth as much as possible the difference between the costs per seat, but this time the leveling process will force the highest price paid to be as small as possible (function (10)). Due to this property Smallest Divisors will tend to include all parties in representation. Just like Equal Proportions, this method is not currently adopted in Europe, although it has been thoroughly discussed in the United States apportionment history. It should be clear by now that each formula embodies a measure of disproportionality or unfairness and, regardless of the particular vote distribution, it always performs its best with respect to the corresponding measure. This is the reason why the widely spread practice in political science of testing the proportionality of the different methods on the basis of one of classic disproportionality indexes is a questionable, if not methodologically erroneous, procedure.[7] In fact, the choice of the index is automatically biased in favor of the proportional method which minimizes it. If the index is, for example, the sum of the absolute deviations between the number of seats assigned to each party and their exact quota, divided by the number of

[7]The analysis of proportionality is still part of an open and often controversial debate; see, for example, Cox and Shugart (1991), Fry and McLean (1991), Gallagher (1991), and Lijphart (1994).

competing parties,

$$I = \frac{\sum_{i=1}^{n} \left| s_i - v_i \frac{S}{P} \right|}{n},$$

then the Largest Remainders method will always be the most proportional, whatever the particular vote distribution. The same is true for the Rae disproportionality index, the Loosemore–Hanby index, and the least squares index (recalled in Table 6.4), which are all multiples or simple transformations of criterion (6), for $\rho = 1$ or $\rho = 2$.

Table 6.4. *Traditional disproportionality indexes.*

Rae index	$\frac{1}{n}\sum_{i=1}^{n}\left\|\frac{v_i}{P} - \frac{s_i}{S}\right\|$
Loosemore–Hanby index	$\frac{1}{2}\sum_{i=1}^{n}\left\|\frac{v_i}{P} - \frac{s_i}{S}\right\|$
Least squares index	$\sqrt{\frac{1}{2}\sum_{i=1}^{n}\left(\frac{v_i}{P} - \frac{s_i}{S}\right)^2}$

The methodological approach we suggest using to analyze the performance of different proportional formulas consists of evaluating disproportionality with respect to several indexes. This has two advantages: first, it will avoid the biased results and, second, a multiple objective approach will lead to useful information on the robustness of these methods. In our opinion, the robustness of a method with respect to different proportionality measures should be considered in itself as a criterion to evaluate the degree of proportionality. A method is *robust* if it simultaneously yields good values for different proportionality indexes in a sufficiently large amount of cases.

A simulation experiment has been carried out [Pennisi, 1997] analyzing 500 random but realistic cases, classified on the basis of the number of competing parties and the degree of vote fractionalization. Eight traditional divisor methods have been tested (Largest Remainders, Droop, Imperiali, d'Hondt, Sainte-Laguë, Equal Proportions, Smallest Divisors, Harmonic Mean) with respect to a set of very different disproportionality indexes (including Gini concentration index, the entropy, and the L_ρ-norms based on the deviation between the shares of representation of single parties and the ratio $\frac{S}{P}$). The results show that, while no method uniformly dominates the others in terms of proportionality, the Largest Remainders method and Sainte-Laguë tend to be robust with respect to the set of indexes chosen.

Before continuing our discussion, notice that all of the unfairness indicators listed in Table 6.1 can be considered as a reasonable way to measure and formulate the concept of proportionality. However, the solution to the proportional representation problem changes depending on which of these criteria is chosen. All of these solutions can be considered "proportional" in spite of the fact that a particular solution may be advantageous for a specific party. The choice of a proportional electoral system should also be based on the choice of an unfairness measure. The difficulty, of course, is to identify a fairness criterion on which everyone agrees and in particular on which the politicians operating an electoral reform will agree.

As mentioned above, the integer optimization approach we have suggested to identify such criteria can also be used to design new electoral systems on the basis of totally different measures of disproportionality. In our opinion, even when analyzing some of the more recent results in the field of proportional apportionment, this possibility is never emphasized enough. For example, Ernst (1994) proves that if the method of Sainte-Laguë produces no quota violations, then such a method minimizes the function $\sum_{i=1}^{n} |\frac{s_i}{v_i} - \frac{S}{P}|$ among all apportionments that do not violate lower and upper quotas. Now, such a function has a strong intuitive appeal because it is a classic measure of the total distance between the actual amount of representation $\frac{s_i}{v_i}$ each party receives and the amount $\frac{S}{P}$ it should receive in a fair situation. Since $\sum_{i=1}^{n} |\frac{s_i}{v_i} - \frac{S}{P}|$ seems to be considered a reasonable objective in order to guarantee a fair seat assignment, why not design a method that *always* minimizes it if this is possible? In fact, the function is separable and convex, so according to Theorem 6.1, a greedy procedure will do. We need only to consider its discrete derivative:

$$\rho_i(s_i) = \left|\frac{s_i+1}{v_i} - \frac{S}{P}\right| - \left|\frac{s_i}{v_i} - \frac{S}{P}\right| = \begin{cases} \frac{1}{v_i} & \text{if } s_i \geq v_i\frac{S}{P}, \\ \frac{2s_i+1}{v_i} - 2\frac{S}{P} & \text{if } v_i\frac{S}{P} - 1 \leq s_i \leq v_i\frac{S}{P}, \\ -\frac{1}{v_i} & \text{if } s_i \leq v_i\frac{S}{P} - 1 \end{cases}$$

for every $i = 1, 2, \ldots, n$, and to assign the S seats one at a time to the party i^* for which $\rho_{i^*}(s_{i^*}) = \min\{\rho_i(s_i) : i = 1, 2, \ldots, n\}$.

Of course, many other reasonable objective functions could be tested and the corresponding optimization algorithms designed, allowing us to develop new proportional electoral methods. Electoral formulas minimizing entropy, the Gini concentration index, standard deviation, and other common indexes of inequality could be just as good as other consolidated methods, as will be further discussed in Chapter 7. Moreover, not only can algorithms minimizing other unfairness measures be found, but adding specific constraints to the optimization problem (P2), we can force seat assignments to respect appealing properties (such as some of those mentioned in section 5.2) in addition to minimizing some suitably chosen "distance" between the actual allocation and the ideal one.

6.5 The Complexity of an Electoral Formula

The ideal electoral formula is both easy to understand and easy to compute. The latter requirement was a technical necessity in the old times when computation was done by hand. The development of computers and information technology nowadays allows the use of more complicated methods and can provide faster solutions. Easy understanding is a much more intrinsic requirement, since in a truly democratic system each elector should have a clear perception of the consequences of his/her vote.

The complexity of the seat allocation method is a measure of the effort that is necessary to determine its result. By viewing these methods as algorithms, an appropriate measure is already at hand. In fact, according to a standard definition in computer science, the *computational complexity* of an algorithm is defined as the number of elementary operations (addition, subtraction, multiplication, division, pairwise comparison, assignment, etc.) that must be performed in the worst case to obtain the output of the algorithm for a fixed size of the problem. The number of elementary operations is computed as a function of the size of the input of the problem.

A similar approach can also be found in Gottinger (1987) who is interested in building a machine capable of simulating the strategies of each voter, that is, their individual preferences. In such a case, the entire society can be represented by a *social choice machine*, which produces the decision of social choice as an output after receiving all the individual preferences. The idea is very similar to that of the *Turing machine* in computational complexity. The complexity of any social choice function is therefore equivalent to that of the "shortest" program on such a machine that can simulate the social choice machine. The input of a seat allocation method is determined by the number of parties (or lists or candidates), the votes received by the parties, and the total number of seats available. The output is a partition of the seats among the parties. The size of the problem is therefore characterized by n and S. It still remains to explain what we mean by *worst case*. Given two problems with the same size (i.e., the same values for both n and S) it is possible that the number of operations needed to solve one problem is larger that the one needed to solve the other, even though the formula used is the same. This just depends on the particular vote distribution. For example, divisor methods need parties to be ordered according to the ratios $\frac{v_i}{d(s_i)}$ at each step. If they are already ordered correctly, the number of pairwise comparisons needed to order them at each step will be smaller than in any other case. The measure we adopt for complexity is based on *pessimistic expectation*: it counts the maximum number of elementary operations that occur to solve any problem of fixed size.

Since there may also be different ways of implementing the same method, and this could itself cause different measures of complexity, the definition we refer to states that the complexity of an electoral formula is equal to the complexity of the most efficient algorithm that can implement it. Denote by $C_A(L)$ the computational complexity of algorithm A for a problem instance in which L bits of memory are needed to store all input data. Let Ω be the set of algorithms that may be used to implement a given electoral formula, F. The complexity of the electoral formula F is defined as follows:

$$C_F(L) = \min C_A(L), \quad A \in \Omega.$$

In most cases it would be much too difficult to count the number of elementary operations exactly. It will suffice to compute an upper bound on the number of elementary operations, given by a simple function of L, which usually depends on the number n of parties and the number S of seats.

One word of caution is needed. In the common practice of computational complexity estimation, usually it will be enough to know that the running time of a certain allocation method is, say, $O(nS)$: in Landau's notation, this means that there are a positive constant c and two integers $\bar{n}$ and $\bar{S}$ such that the number of elementary operations is at most cnS for all $n > \bar{n}$ and $S > \bar{S}$.

Since the main concern is the rate of growth of the running time as the size parameters n and S increase, one is usually not interested in the actual values of $\bar{n}$, $\bar{S}$, and c. However, if one had, say, $\bar{n} = 1,000$ and $\bar{S} = 100,000$, then the above upper bound would be useless, since in all countries of the world the number of parties n does not exceed 100 and the number of seats S does not exceed a few thousand. The value c is also of importance. Usually, a method whose running time is $O(nS)$ is deemed to be better than a method with running time $O(n^2S^2)$. But if this means that the first running time is at most $10^9\, nS$ when $n > 2$ and $S > 3$, while the second one is $10^2n^2S^2$ when $n > 2$ and $S > 3$, then surely one would prefer the second method. In fact, in all cases of interest, one has, say, $n \leq 100$ and $S \leq 3,000$, implying that $10^2n^2S^2 < 10^9nS$.

Finally, knowing that the running time of an allocation method is bounded above by $c \times g(n, S)$ for a suitable function g when $n > \bar{n}$ and $S > \bar{S}$ is of practical use only when the following two conditions are met:

(1) the integers $\bar{n}$ and $\bar{S}$ are smaller than the actual values of n and S, respectively, in all cases under consideration;

(2) the constant c is relatively small.

Subject to these conditions, the complexity of an actual formula can be estimated reasonably well.

Chapter 7

Rewarded and Punished Parties

The point made in Chapter 6 is that the solution to the problem of assigning seats proportionally to votes (or to population) is not unique. Electoral formulas are methods designed to minimize the distance between vote and seat shares, but the problem is that such a distance can be measured in many different ways. Each formula corresponds to a specific measure and yields an optimal result for the proportional representation problem where such a measure is chosen as the objective. Nevertheless, once the seats have been assigned, some parties receive a number of seats that is larger than the amount they should have obtained and other parties receive a bit less than what they deserve since the integral nature of the seats usually makes perfectly fair seat assignment impossible.

In this chapter we try to analyze the inequality that arises between parties when the seats are assigned. A new way to characterize proportional seat assignment methods based on the concept of *majorization* is presented, pointing out the similarities between the measurement of disproportionality and the measurement of income inequality. Our main purpose is to introduce the class of Schur-convex functions as an interesting class of fairness measures.

7.1 Handling Over- and Underrepresentation

Given a seat distribution $\mathbf{s}$, party i is *overrepresented* when the ratio between seats and votes party i receives is larger than the ratio between total seats and total votes:

$$\frac{s_i}{v_i} > \frac{S}{P},$$

or, in other words, when the price party i has paid for each seat is smaller than the real cost per seat,

$$\frac{v_i}{s_i} < \frac{P}{S}.$$

On the other hand, party i is *underrepresented* when the opposite inequalities hold. Obviously, fair assignment of seats implies that the ratio $\frac{s_i}{v_i}$ produces the same value for every i and, in particular, it implies that there are no over- nor underrepresentations. In fact,

$$\frac{s_i}{v_i} = \text{constant } \forall i \Rightarrow \frac{s_i}{v_i} = \frac{S}{P} \quad \forall i.$$

We can view the ratio between seats assigned and votes received by a given party as the *quality of the representation* of such party in the system. On the other hand, the ratio between total seats and total votes can be thought of as the scale factor adopted to convert the weight of each party in the nation to its weight in Parliament. A party is perfectly represented only when its seat-to-vote ratio is equal to the fixed scale factor $\frac{S}{P}$. Therefore, an idea of the global quality of representation obtained with an electoral procedure can be derived by the following vector, called *representation vector*:

$$\boldsymbol{y} = (y_1, y_2, \ldots, y_n),$$

where

$$y_i = \frac{s_i}{v_i}, \quad i = 1, 2, \ldots, n.$$

The single elements of the vector allow us to identify who has been rewarded and who has been punished by the current seat assignment. The following question naturally arises: can the quality of the representation or, equivalently, the fairness of the seat distribution be improved by shifting some seats from overrepresented to underrepresented parties? The answer is obviously connected to the index used to measure such inequality.

At the beginning of this century, a very similar debate took place among the pioneers of *welfare economics*. Well-known researchers such as Lorenz, Pigou, and Dalton started to be interested in the development of a method to compare different income or wealth distributions. The core of the discussion was in fact how to determine a correct measure of inequality and to identify which income distributions were "more equal" (fair) than others. The concept of "equal distribution" of any kind of resource among a set of individuals can intuitively be defined as the distribution that is the least concentrated possible.

The first attempt of a methodical analysis of income inequality is due to Lorenz (1905), who formally defines the concept of "fair" wealth distributions and compares them by using the method now commonly known as *Lorenz curves*. Lorenz curves represent the cumulative shares of wealth corresponding to increasing amounts of population. If $\boldsymbol{y} = (y_1, y_2, \ldots, y_n)$ is the wealth distribution corresponding to n persons and these persons are ordered from the poorest to the richest, then the corresponding Lorenz curve passes through the points (h, T_h) for $h = 0, 1, \ldots, n$ where

$$T_h = \frac{\sum_{i=1}^{h} y_i}{\sum_{i=1}^{n} y_i} = \frac{\sum_{i=1}^{h} y_i}{T},$$

and T is the total amount of wealth or income. If wealth is distributed in the fairest manner among the n individuals, the points above are totally aligned and the curve is actually a straight line. Otherwise, we will get a convex curve lying under this straight line and with the same end points. The more the curve is bent, the more the distribution is concentrated or unequal. In Figure 7.1, for example, the wealth distribution corresponding to curve B is less concentrated or fairer than that corresponding to curve C, but the fairest distribution is always represented by curve A.

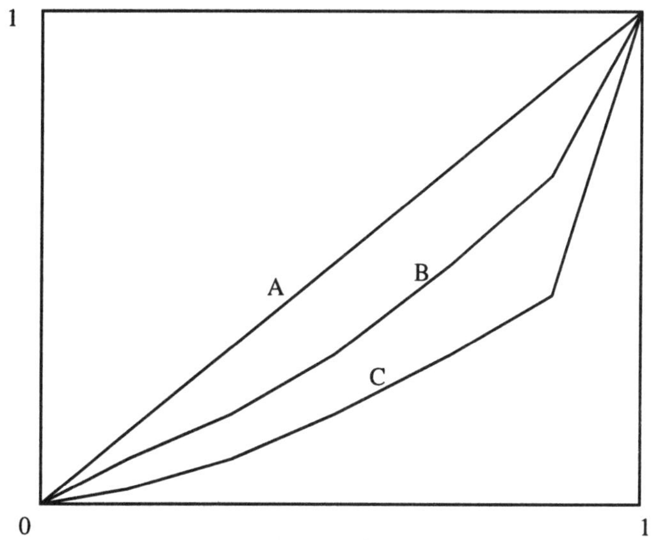

Fig. 7.1. *Lorenz curves.*

This kind of representation also suggests a way to handle the problem of determining the fairest seat distribution. Our intention is always that of finding a vector of seats $\boldsymbol{s}$ that will guarantee that the corresponding vector of representation $\boldsymbol{y}$ is the less concentrated possible.

The representation made by Lorenz curves immediately shows that, given the total amount T of some resource, distribution $\boldsymbol{a} = (a_1, a_2, \ldots, a_n)$ is smoother or less concentrated than $\boldsymbol{b} = (b_1, b_2, \ldots, b_n)$ if, for every $h = 1, 2, \ldots, n$, the sum of the h largest elements in $\boldsymbol{a}$ is always smaller than or equal to the sum of the largest h elements in $\boldsymbol{b}$ and equality holds when $h = n$:

$$\begin{aligned} \sum_{i=1}^{h} a_{[i]} &\leq \sum_{i=1}^{h} b_{[i]}, \qquad h = 1, 2, \ldots, n-1, \\ \sum_{i=1}^{n} a_{[i]} &= \sum_{i=1}^{n} b_{[i]} = T, \end{aligned} \tag{7.1}$$

where $a_{[1]}, a_{[2]}, \ldots, a_{[n]}$ are the components of the vector $\boldsymbol{a}$ disposed in nonincreasing order. The inequalities in (7.1) are one of the many ways to formulate the concept of *majorization* [Hardy, Littlewood, and Polya, 1929; Marshall and Olkin, 1979]. In this case, it can also be said that vector $\boldsymbol{a}$ is *majorized* by vector $\boldsymbol{b}$.

Basically, majorization places at the bottom those vectors whose components tend to be as equal as possible. For example, consider five individuals who globally own \$100. The most concentrated situation is that in which one of the five owns the total amount and the other four own nothing, thus defining wealth distributions such as (\$100,0,0,0,0) or (0,0,\$100,0,0), or any other permutation of these elements. The less concentrated situation is that in which each person owns an equal amount of dollars, i.e., (\$20,\$20,\$20,\$20,\$20). Consider a vector $\boldsymbol{y}^* = (y^*, y^*, \ldots, y^*)$ where each element is equal to the average value

$$y^* = \frac{\sum_{i=1}^{n} y_i}{n}.$$

The vector $\boldsymbol{y}^*$ is majorized by any other vector $\boldsymbol{y} = (y_1, y_2, \ldots, y_n)$ with the same number of components and the same total sum, $\sum_{i=1}^{n} y_1 = ny^*$.

Several years after Lorenz, Dalton (1920) became interested in the same problem but from another point of view. In particular, the idea Dalton brought forth is condensed in what he called the *principle of transfers*:

> If there are only two income-receivers and a transfer of income takes place from the richer to the poorer, inequality is diminished. There is indeed a limiting condition. The transfer must not be so large as to more than reverse the relative positions of the two income receivers and it will produce its maximum result, that is to say, create equality, when it is half the difference between the two incomes. And we may safely go further and say that, however great the number of income receivers, any transfer between any two of them, or, in general, any series of such transfers, subject to the above condition, will diminish inequality. It is possible that, in comparing two distributions, in which both the total income and the number of income-receivers are the same, we may see that one might be evolved from the other by means of a series of transfers of this kind. In such case we could say with certainty that the inequality of one was less than that of the other.

Consider a vector $\boldsymbol{y} = (y_1, y_2, \ldots, y_n)$ representing the allocation of a fixed amount of resources into n parts. Let (h, l) be a pair in which h is "richer" than l, meaning that $y_h > y_l$. If a positive amount of the resource, Δ, is transferred from h to l, the total inequality contained in the vector decreases provided that

$$y_h - \Delta \geq y_l + \Delta,$$

that is,

$$y_h - y_l \geq 2\Delta, \tag{7.2}$$

or, in other terms, that the relative position of h and l is not reversed by the transfer.

The typical situation is shown in Figure 7.2. This kind of transfer is *symmetric* (or *zero sum*) since h loses exactly what l gains.

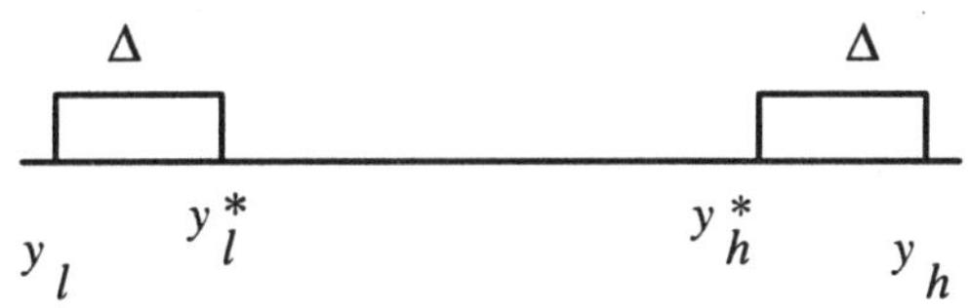

Fig. 7.2. *Symmetric transfer.*

The inequality in $\boldsymbol{y}$ can be reduced by performing successive transfers of this kind from a richer to a poorer individual.

Let us try to use the same rationale where seat allocations are concerned. Rich individuals correspond to overrepresented parties while poor individuals correspond to underrepresented parties. The main difference is that over- and underrepresentation is related to the comparison between the amount of representation allocated to parties and the fixed quota $\frac{S}{P}$ rather than to the comparison of seats between single parties. Therefore, if we want to allocate a fair number of seats to each party, we must choose the seat vector $\boldsymbol{s} = (s_1, s_2, \ldots, s_n)$, which makes the corresponding representation vector $(\frac{s_1}{v_1}, \frac{s_2}{v_2}, \ldots, \frac{s_n}{v_n})$ as leveled as possible. Since the principle of transfers provides a sufficient condition for the improvement of a distribution vector, given a representation vector, we should check whether it is possible to perform a similar transfer from overrepresented to underrepresented parties. The main difficulty is that the transfer between parties is constrained to the fact that the seats are integers and this additional condition must be considered with (7.2).

Notice that this kind of approach is mentioned *in fieri* by Huntington (1921), who analyzes the inequality in representation between pairs of parties.

From this viewpoint, the proportional representation problem basically consists of finding the most equally distributed vector of party representation by performing consecutive transfers of seats, considering both the constraint on the maximum amount that can be transferred and the constraint on the integral nature of these transfers. This case is more complicated than the one described above because the transfers are performed on the vector of seats, but the implications we are interested in are measured on the vector of party representation.

In more detail, the symmetric transfers performed on the vector of seats correspond to *asymmetric transfers* on the representation vector. Suppose a single seat is transferred from party h to party l. The representation ratios of these two parties will be modified from

$$y_h = \frac{s_h}{v_h} \text{ and } y_l = \frac{s_l}{v_l}$$

to

$$y_h^* = \frac{s_h - 1}{v_h} \text{ and } y_l^* = \frac{s_l + 1}{v_l},$$

respectively. Notice that the gain of party l (which is equal to $\frac{1}{v_l}$) is different from the loss party h is affected by (i.e., $\frac{1}{v_h}$), unless the number of votes cast for the two parties, v_h and v_l, is the same. Also notice that while the total amount of resource in the seat vector is unchanged by the transfers (since the sum of the seats is always equal to the total number of available seats S), the total sum in the party representation vector changes. Figure 7.3 illustrates an example of an asymmetric transfer on the representation vector.

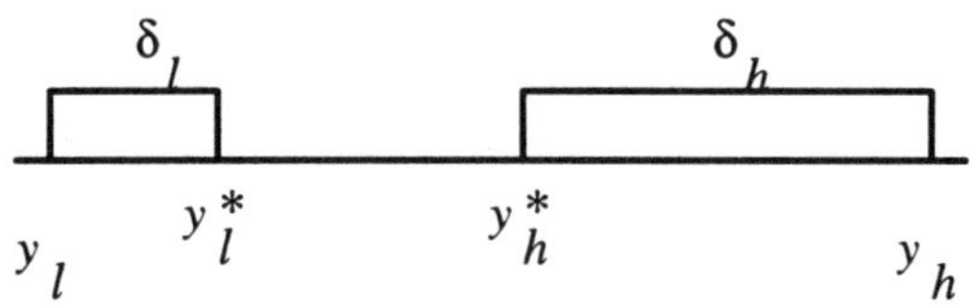

Fig. 7.3. *Asymmetric transfer.*

More formally, the simple transfer of a seat between two parties corresponds to a δ-transfer in terms of representation. Let us define a *δ-transfer* between h and l as an asymmetric operation such that

$$\begin{aligned} y_h^* &= y_h - \delta_h, \\ y_l^* &= y_l + \delta_l, \\ y_i^* &= y_i \quad \forall i \neq h, l, \end{aligned}$$

under the following constraints:

$$\begin{aligned} y_h - y_l \geq \delta_h + \delta_l, \\ \delta_h, \delta_l > 0. \end{aligned}$$

Additional conditions must hold since the transfer actually occurs in terms of seats, which must be integers and which must sum up to S in total. In general, this means that

$$\delta_h = s\frac{1}{v_h} \text{ and } \delta_l = \delta_h \frac{v_h}{v_l},$$

where s is the integer number of seats transferred from h to l. This suffices to guarantee that the number of seats is an integer and the total number of seats available is unchanged.

7.2 Measures of Inequality between Parties and Exchange Algorithms

The theory of majorization can be a useful tool in order to better understand and evaluate the inequality arising in proportional representation systems. In this context an important role is assigned to the class of Schur-convex functions. In fact, Schur-convex functions preserve the partial order induced by majorization and are therefore particularly suitable to measure the degree of inequality contained in a vector.

Typical examples of Schur-convex functions are

$$\varphi^{(1)} = \frac{1}{n}\sum_i (x_i - \bar{x}),$$

where $\bar{x} = \frac{1}{n}\sum_{i=1}^n x_i$ is the arithmetic mean of $x_1, x_2, \ldots, x_n$;

$$\begin{aligned} \varphi^{(2)} &= \sum_i x_i^2; \\ \varphi^{(3)} &= \sum_i \sum_j |x_i - x_j|. \end{aligned}$$

Schur-convex functions take smaller values when their arguments $x_1, x_2, \ldots, x_n$ are closer to each other. Looking back at Figure 7.1, any Schur-convex function will reach its minimum value for curve A (which corresponds to a perfectly equal distribution) and will take increasing values when curves A, B, and C are considered. This is the reason why they are particularly fit to measure and to compare the degree of fairness embodied in a distribution of resources.

A precise definition of Schur-convexity will be now given. A square $n \times n$ matrix B is *doubly stochastic* if all its entries b_{ij} are nonnegative reals and both its row and column sums are equal to 1; that is,

$$\begin{aligned} & b_{ij} \geq 0 \quad \forall i,j, \\ & \sum_{j=1}^n b_{ij} = 1 \quad \forall i, \\ & \sum_{i=1}^n b_{ij} = 1 \quad \forall j. \end{aligned}$$

When a vector $\boldsymbol{x}$ is premultiplied by a doubly stochastic matrix, the sum of the components of the new vector remains unchanged.

Let $X \subseteq \Re^n$ be such that $\boldsymbol{x} \in X \Rightarrow B\boldsymbol{x} \in X$ for every doubly stochastic matrix B. A real-valued function $\varphi(\boldsymbol{x})$ defined in X is *Schur-convex* if

$$\varphi(\boldsymbol{x}) \geq \varphi(B\boldsymbol{x}) \quad \forall B, \forall \boldsymbol{x} \in X.$$

It can be shown that, for any two vectors $\boldsymbol{x}, \boldsymbol{x}' \in \Re^n$, the following three properties are equivalent:

(1) $\boldsymbol{x}'$ is majorized by $\boldsymbol{x}$;

(2) $\boldsymbol{x}'$ can be obtained from $\boldsymbol{x}$ by a finite sequence of transfers,

(3) $\boldsymbol{x}' = B\boldsymbol{x}$ for some doubly stochastic matrix B.

Recall the optimization problem formulated in Chapter 6 and the objective functions associated with different proportional representation methods. Since such objective functions must be measures of unfairness or inequality, it seems natural to refer to the class of Schur-convex functions [Pennisi, 1996]. In particular, if the measure of inequality is Schur-convex, we can design a proportional allocation method by solving the following optimization problem:

$$\text{minimize } \varphi\left(\frac{s_1}{v_1}, \frac{s_2}{v_2}, \ldots, \frac{s_n}{v_n}\right),$$

$$\sum_{i=1}^{n} s_i = S,$$

$$s_i \geq 0, \quad \text{integer}, \quad i = 1, 2, \ldots, n,$$

where $\varphi(\boldsymbol{y})$ is a Schur-convex function.

This formulation is consistent with the disproportionality indexes we have already analyzed and their properties. In fact, several functions listed in Table 6.1 are Schur-convex functions of $(\frac{s_1}{v_1}, \frac{s_2}{v_2}, \ldots, \frac{s_n}{v_n})$.

The principle of transfers has a straightforward application in the theory of inequality and it suggests an interesting procedure for proportional representation, too. In fact, given an arbitrary seat allocation satisfying the cardinality constraint $\sum_{i=1}^{n} s_i = S$, by transferring seats from one party to another according to an appropriate *exchange algorithm*, one can design a procedure that minimizes most of the objective functions we are interested in. In fact, an important subclass of Schur-convex functions (including functions (6) to (9) in Table 6.1) is the class of those functions φ such that $\varphi(\boldsymbol{s}) = \sum_{i=1}^{n} g(\frac{s_i}{v_i})$, where g is a convex function. Notice that $\varphi_i(s_i) = g(\frac{s_i}{v_i})$ is a convex function of s_i. Hence, by the results of section 6.3, the greedy algorithm provides a seat allocation that minimizes $\varphi(\boldsymbol{s})$. It will be shown below that the exchange algorithm provides a seat allocation that coincides with at least one (ties are possible) of the optimal solutions produced by the greedy algorithm. Therefore, the exchange algorithm yields an optimal solution as well. Consider a separable function $\varphi(\boldsymbol{s}) = \sum_{i=1}^{n} \varphi_i(s_i)$. Given a seat allocation $\boldsymbol{s}$, we are interested in finding a general rule to establish whether a seat transfer from party i to party j is convenient when trying to minimize $\varphi(\boldsymbol{s})$.

Let

$$\varphi(\boldsymbol{s}') = \varphi(s_1, \ldots, s_i - 1, \ldots, s_j + 1, \ldots, s_n)$$

be the value of the function after the transfer of a seat between i and j. The seat allocation after the transfer is "better" than the one before the transfer when it yields a smaller value for our objective, i.e., when $\varphi(\boldsymbol{s}') < \varphi(\boldsymbol{s})$.

The transfer of a seat from party i to party j will be said to be *convenient* or *profitable* if $\Delta_{ij} \equiv \varphi(\boldsymbol{s}') - \varphi(\boldsymbol{s}) < 0$.

A seat allocation $\boldsymbol{s}$ can be defined *exchange stable* with respect to function $\varphi(\boldsymbol{s})$ if there is no pair of parties for which a seat transfer is profitable.

Recall from section 6.3 that any separable function φ can be written as the sum of a constant value and a function of its discrete derivatives and, more precisely, $\varphi(\boldsymbol{s}) = K + h(\boldsymbol{s})$. Therefore, a solution $\boldsymbol{s}^*$ minimizing $h(\boldsymbol{s})$ also minimizes $\varphi(\boldsymbol{s})$.

LEMMA 7.1. *Let $\varphi(\boldsymbol{s})$ be a separable function with nondecreasing discrete derivative ρ_i. For a seat allocation $\boldsymbol{s}^*$ the following statements are equivalent:*

(1) *$\boldsymbol{s}^*$ is optimal, i.e., it minimizes $\varphi(\boldsymbol{s})$ over all seat allocations $\boldsymbol{s}$ such that $\sum_{i=1}^n s_i = S$;*

(2) *$\boldsymbol{s}^*$ is exchange stable;*

(3) *for all parties i and j, with $s_i \geq 1$, one has*

$$\rho_j(s_j^*) \geq \rho_i(s_i^* - 1); \tag{7.3}$$

(4) *$h(\boldsymbol{s}^*) = l_S$, where l_S is defined as in section* 6.3.

Proof.

(1) $\Rightarrow$ (2): Obvious.

(2) $\Rightarrow$ (3): Let $s_i \geq 1$, and let $\boldsymbol{s}^{**}$ be the seat allocation obtained by the transfer of one seat from party i to party j. Since $\boldsymbol{s}^*$ is exchange stable, one has

$$0 \leq \varphi(\boldsymbol{s}^{**}) - \varphi(\boldsymbol{s}^*) = \varphi_j({s_j}^* + 1) + \varphi_i({s_i}^* - 1) - \varphi_j({s_j}^*) - \varphi_i({s_i}^*)$$
$$= \rho_j({s_j}^*) - \rho_i({s_i}^* - 1),$$

implying (7.3).

(3) $\Rightarrow$ (4): Follows from Lemmas 6.1 and 6.3.

(4) $\Rightarrow$ (1): Follows from Lemma 6.2. □

Condition (3) will be called the *stability condition*. An *exchange algorithm* consists of performing a sequence of seat transfers between parties until the stability condition holds for every pair of parties.

THEOREM 7.1. *If φ is a separable function and its discrete derivatives ρ_i are nondecreasing, any exchange algorithm outputs an optimal allocation $\boldsymbol{s}^*$.*

Proof. An exchange algorithm provides an exchange stable solution $\boldsymbol{s}^*$; therefore, by Lemma 7.1 $\boldsymbol{s}^*$ is optimal. □

Let us call *unethical* an exchange of one seat from some underrepresented party to some overrepresented one (any other exchange is called *ethical*). Using the expressions of the discrete derivatives given in Table 6.2, one can easily check that unethical exchanges are never profitable for the corresponding unfairness measures, as it should be expected. Hence, all such measures can be minimized via a sequence of *ethical exchanges*. Therefore, over- and underrepresentation play a fundamental role in designing operational procedures to reduce unfairness.

Chapter 8

Mixed Electoral Systems: Choosing the Best Mix

Mixed (or hybrid) electoral systems are usually considered an attempt to include both the advantages in terms of representation and those in terms of government stability attributed to the proportional and majoritarian methods, respectively. Such systems are explicitly designed to obtain "*the best of both worlds*" [Lijphart, 1986]. Although they are today in force in no less than 25 countries, including about 16% of the world population (Table 8.1), they are treated in the literature as a secondary subject or as an appendix of the proportional or plurality system, depending on which has the major role in the seat allocation.

In fact, democratic representation has historically evolved through the concept of one man, one vote. This can be carried out either by adopting a plurality electoral formula with fair (proportional) apportionment or by directly adopting a proportional electoral formula. While the different variants of both proportional and majoritarian formulas have prospered since the very beginning, the use of a mixture of these systems has appeared only in the last century (the first example is Denmark in 1915; see Elklit, 1992). The idea underlying such seat allocation methods is not always clear to the voter and, perhaps, not even to political engineers.

Table 8.1. *Classification of electoral systems in force in* 1996.

Electoral system	*No. of countries*
Plurality systems	59
Proportional representation	56
Mixed systems	25
Majority systems	25

8.1 General Description of Mixed Electoral Systems

A systematic classification of hybrid methods is not trivial. In fact, "pure" electoral systems hardly exist in practice due to a number of special features that are usually added to the simple vote-to-seat formula such as provisions for minority protection, exclusion thresholds, majority bonuses, etc. The difficulty of drawing boundaries between the different kinds of methods is reflected by the confusion often made while mentioning hybrid systems, to the point that sometimes methods that are essentially majority systems but adopt multimember districts are called mixed.

In a recent classification [Blais and Massicotte, 1996] three types of hybrid methods are identified and defined on the basis of the way the majoritarian and proportional formulas are blended. The blending alternatives are

- coexistence,
- combination,
- correction.

When the electoral system is basically the first-past-the-post method but, in restricted areas of the territory, a few districts adopt a proportional method, the system will be classified hybrid by *coexistence*. Examples of this type usually occur in countries where ethnic or cultural minorities must be protected. On the other hand, mixed systems in which the proportional seats are allocated independent of the outcome under the plurality or majority rule are called hybrid by *combination* (such as the system currently used in Japan and Russia). Finally, the term *correction* is used when the proportional seats are distributed so as to compensate for the distortion generated under plurality or majority (examples are given by Denmark in 1915 and by Germany, Italy, and Hungary today). Maybe hybrid systems by correction are the most complicated but interesting case. This class of methods has many variants since the proportional seats are not necessarily distributed on the basis of *all* the votes cast. In Italy and in Hungary, for example, proportional seats are distributed only on the basis of the votes cast for defeated single-member candidates and considering the *winning margin of votes* for the others. The rationale behind such a rule is that under the first-past-the-post method the votes cast for defeated candidates as well as the winning candidate's margin of victory (in excess of one) are "wasted" and ought to be recycled in order to produce fairer representation. The other votes (those that are necessary to the winner in each district in order to win) must not be considered because they have already been "used."

Throughout this chapter we will only consider mixed systems that are hybrids by correction, where a fixed fraction of seats is assigned by a majority-type rule and the remaining fraction by a proportional formula but in a dependent manner. Some concrete examples of these systems are listed in Table 8.2.

A huge number of mixed systems can be designed by modifying not only the majority/proportional ratio but also the particular majority and proportional method taken into consideration. Some blends are more famous than others. For example, the mix ratio adopted in the German system and in Russia is 50/50 between plurality and the Hare method; in Italy (Senate) it is 75/25 between the first-past-the-post and d'Hondt methods; in New Zealand 50/50 between the first-past-the-post and Sainte-Laguë methods, while in Japan it is 60/40 between the first-past-the-post and d'Hondt methods, and in Albania 70/30 between double ballot and d'Hondt methods.

Table 8.2. *Countries that adopt mixed systems* (1996).

Africa	Guinea, Senegal, Niger, Cameroon, Tunisia, Seychelles
Asia	Japan, Taiwan, Azerbaidjan, Georgia
North America	Mexico, Panama, Guatemala
South America	Venezuela, Ecuador, Bolivia
Oceania	New Zealand
Europe	Germany, Italy, Andorra
New European Democracies	Russia, Hungary, Croatia, Lithuania, Albania

We define *dosage* of a mixed system the fraction α of seats that is to be distributed with a majoritarian formula. Pure methods can be thought of as extreme cases: the dosage $\alpha = 1$ corresponds to the specific majoritarian formula and $\alpha = 0$ to the proportional formula.

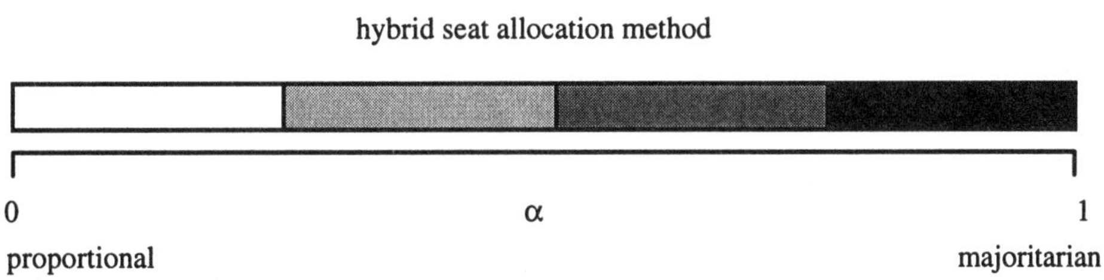

Fig. 8.1. *The range of mixed systems.*

The *detraction rule*, which defines the number of ballots and the votes to be used in the proportional allocation phase, is an essential feature in order to define mixed systems. The aim of such a rule is to enhance the opportunity of gaining a seat for the parties or coalition that was not able to win in the single-member districts. In fact, if no detraction is made, the parties that have already won a seat are favored with respect to those that have been excluded from the majoritarian representation phase. Therefore, the minimum number of votes necessary to win in the single-member district (which is a different number in each district depending on the votes cast for the other parties) is subtracted from the total votes obtained by each party in all the districts where such a party has won. The minimum number of votes necessary to win in a single-member district (the *winning margin* of votes) is equal to the number of votes cast for the second biggest party plus one.

There are two main types of detraction rules. When the electoral system requires only one ballot, it will be a *detraction from the total*, just as described above. If several parties compete in the elections together in a unique coalition, it is generally required that the coalition be the same in all districts. Let $I = \{1, 2, \ldots, n_R\}$ be the set of parties (or coalitions) competing in region R, $J = \{1, 2, \ldots, k_R\}$ be the set of single-member districts in region R, and $J(i) \subseteq J$ be the set of districts in which the candidate supported by party i is the winner of the majority seat. Then let v_{ij} be the number of votes cast for party i in district j and let $\bar{v}_i$ be the net regional vote for party i. The majority seat is won in each district j by the candidate of that party i^* for which the following holds:

$$v_{i^*j} = \max_{l \leq i \leq n_R} v_{ij}.$$

When the total detraction rule is adopted, the net regional vote for party i will be determined as follows:

$$\bar{v}_i = \sum_{j=1}^{k_R} v_{ij} - \sum_{j \in J(i)} v_{ij}.$$

On the other hand, when the electoral system requires two ballots, one for the majority or plurality phase and the other for the proportional seat allocation, a *pro-quota detraction rule* can be adopted. In this case, the votes subtracted from the winner's coalition in each district (determined by the first ballot) are distributed to the members of the coalition on a proportional basis with respect to the votes cast for the single parties in the second ballot. Let K_j be the set of all candidates in single-member district j. The set of parties supporting a candidate $h \in K_j$ is denoted by $C(h) \subset I$. It can be a single party, $|C(h)| = 1$, or a coalition; in any case, the following conditions hold:

$$\bigcup_{h \in K_j} C(h) \subseteq I,$$

$$C(h) \cap C(l) = \emptyset \quad \forall h, l \in K_j;$$

that is, the family of sets $\{C(h) : h \in K_j\}$ is a *packing* of I. Since the voters cast two ballots, let $v_{hj}^{(1)}$ be the number of votes cast for candidate h in district j on the first ballot and let $v_{ij}^{(2)}$ be the number of votes cast for party i in district j on the second ballot. As before, the majority seat is won in district j by candidate h^* such that

$$v_{h^*j}^{(1)} = \max_{h \in K_j} v_{hj}^{(1)},$$

but this time the net regional vote for party i is determined as the sum of the net districts votes $\bar{v}_{ij}$, which are in turn defined on the basis of the *pro-quota* detraction rule as follows:

$$\bar{v}_{ij} = v_{ij}^{(2)} - w_{ij},$$

where

$$w_{ij} = \begin{cases} 0 \text{ if } i \notin C(h^*), \\ \min\left(v_{ij}^{(2)}, \left\lfloor \dfrac{v_{ij}^{(2)}}{\sum\limits_{l \in C(h^*)} v_{lj}^{(2)}} \cdot \left(\max\limits_{h \neq h^*} v_{hj}^{(1)} + 1\right) \right\rfloor\right) & \text{if } i \in C(h^*). \end{cases}$$

The second case considers the minimum between the *pro-quota* detraction and the actual number of votes the party has received on the second ballot because (clearly) the net district votes per party cannot be negative.

The rationale underlying such a rule is that the parties must pay for the victory of their common candidate a quota which corresponds to their weight in the coalition determined by the votes obtained when competing on their own (second ballot). This kind of subtraction is clearly the most complicated. Moreover, adopting two ballots means that *panachage* is allowed; that is, in the second ballot a voter does not necessarily have to vote for one of the parties belonging to the coalition he voted for in the first ballot.

Hybrid systems must not be considered as the only alternative to correct the disadvantages in terms of government stability that sometimes arise from proportional allocation methods, especially when the number of competing parties is large and they are homogeneously spread on the territory. Another tool that can be used is the variation of the magnitude of the electoral districts. Giannuli (1992) suggests partitioning the country into a big number of small multimember districts and continuing to adopt a proportional method without redistributing seats at national level. In fact, if 300 seats must be allocated and there are 6 districts (of 50 seats each), a party needs only 2% of the votes cast in one of the districts to win a seat. But, if the same number of seats are distributed on the basis of 100 districts (of three seats each), a party needs to gain at least 33.3% of the votes cast in one of the districts to be sure to get a seat. This will usually be enough to eliminate small and medium parties from representation.

Nevertheless, hybrid systems are widely adopted with the motivation described above. But, if one has decided to adopt a hybrid system, the following problem still remains: how should α be chosen? Sometimes the choice reflects the relative strength of those who favor either method. For example, in the Italian *referendum* held on April 18, 1993, 82% of the voters expressed their preference for the first-past-the-post method. The subsequent electoral law stated that 3/4 of the seats should be allocated by the first-past-the-post (the fact that this percentage is slightly smaller than 82% can be explained by the reluctance of some parties to abandon the previous proportional system).

While selecting the parameter α, a good deal of attention should be paid to the ensuing consequences, some of which might be not obvious at all. A bit of mathematics can help understand the effect of mixed allocation methods in terms of the fraction α of seats distributed on a plurality basis. A key question is the following: is it possible to determine an "optimal" mix α^* in a hybrid system in order to achieve a set of well-defined political objectives?

Two partial answers to this question are given in the following sections: the first is a quantification of the minimum loss of seats that certain parties unavoidably undergo when a mixed system is used instead of a proportional system, given that the votes do not change; the second concerns the effect of variations of α on some typical performance indicators concerning government stability, the degree of disproportionality and seat fractionalization, observed by simulating the electoral strategies of a given number of parties.

8.2 Effects on the Electoral Outcome of Small Parties

Let us first deal with the unavoidable seat loss suffered by certain parties when going from a proportional system to a mixed one, assuming that the votes they receive do not change. We specifically refer to those (small or even medium) parties that do not get first-past-the-post seats in any single-member district, either because of their insufficient strength or because of their poor concentration on the territory. We emphasize the fact that this result does not depend on the voting profile: the seat loss predicted by the theorem below is *ineluctable*, whatever the vote is.

To be specific, we shall consider a mixed system whereby a share α ($0 \leq \alpha \leq 1$) of the total number S of seats is assigned by first-past-the-post in single-member districts, while the remaining $(1-\alpha)S$ seats are assigned according to the Largest Remainders method. Furthermore, we shall make the following assumptions.

- There is a single ballot; each candidate runs for a single party—rather than for a coalition—and the votes obtained in the district are attributed (possibly after detraction) to his or her party for the proportional distribution of the remaining $(1-\alpha)S$ seats. Therefore, the possibility of *panachage* is ruled out.

- There is no exclusion threshold; thus, even small parties are admitted to the proportional allocation.

- The single-member districts are grouped into q constituencies (or regions) where the proportional seats are allocated. In each constituency the number of *net votes* is given by the difference between the number of *valid votes* and the number of *detracted votes*. In this case the number of detracted votes is assumed to be equal to the total number of votes obtained in the districts of the constituency by the second best candidates (this rule is a simplification of the total detraction rule).

The above system is a slightly simplified version of the one currently adopted for the election of the Italian Chamber. Let $\omega_i = \frac{v_i}{V}$ be the fraction of votes obtained by party i. If $\omega_i \leq \frac{1}{S}$ then party i will be called *tiny*. We shall make the technical assumption that the party under consideration is not tiny.

Consider now any party i that does not get any seat by the first-past-the-post method. Since such a party is not affected by detraction, the number of votes available for proportional allocation is equal to the total number v_i of its votes. Hence, if N is the overall number of net votes in the q regions, party i gets a number t_i of proportional seats which is equal to the integer part of the quotient $v_i \frac{(1-\alpha)S}{N}$, with the possible addition of one more seat by the largest remainders rule. Clearly, t_i depends on the chosen dosage α.

Denote by s_i the number of seats that would be assigned to party i under the pure Largest Remainders rule (assuming the same vote distribution). The percent loss of seats party i necessarily undergoes when a pure proportional system ($\alpha = 0$) is substituted by a mixed system with dosage α is equal to

$$\lambda_i = 100 \, \frac{s_i - t_i}{s_i}.$$

The following theorem states that the loss λ_i will never drop below a certain threshold that depends on α.

THEOREM 8.1. *When a pure proportional system ($\alpha = 0$) is substituted by a mixed system with dosage α (and the vote distribution does not change), every party i that is not tiny and does not win any single-member seat, whatever the vote, will lose a percentage of seats that is no smaller than*

$$100 \left(1 - \frac{2(1-\alpha)}{(1 - \frac{1}{\omega_i S})} - \frac{1}{(\omega_i S - 1)} \right).$$

Notice that, since party i is not tiny, both denominators in the percentage of seats loss are positive.

Proof. Let v_{1j} and v_{2j} be the number of votes obtained by the best and by the second-best candidate, respectively, in district j. Since $v_{1j} \geq v_{2j}$, the following inequalities hold:

$$2v_{2j} \leq v_{1j} + v_{2j} \leq W_j, \tag{8.1}$$

where W_j is the total number of valid votes in district j.

Let J_R be the set of districts in region R. The total number of valid votes in region R is given by the sum of the valid votes in the districts of J_R:

$$V_R = \sum_{j \in J_R} W_j. \tag{8.2}$$

Let d_R be the sum of votes received by the second-best candidates in J_R, that is,

$$d_R = \sum_{j \in J_R} v_{2j}.$$

Inequalities (8.1) and the definition (8.2) imply that

$$d_R \leq \frac{V_R}{2}. \tag{8.3}$$

Since the number of detracted votes is d_R, the number of net votes in region R is equal to

$$N_R = V_R - d_R.$$

Hence, $N_R \geq \frac{V_R}{2}$ by (8.3). It follows that, if $V = \sum_{R=1}^{q} V_R$ is the total number of valid votes and $N = \sum_{R=1}^{q} N_R$ is the total number of net votes, one must have

$$N \geq \frac{V}{2}. \tag{8.4}$$

Let us now consider the number of seats obtained by individual parties. The number of seats obtained with a total of v_i votes by Largest Remainders is $s_i \geq \frac{v_i}{V}S - 1$, while the number of seats obtained under a mixed system with dosage α, assuming that party i gets no single-member seat, is

$$t_i \leq \frac{v_i}{N}(1-\alpha)S + 1.$$

Hence, the percentage loss party i undergoes when a mixed system is used instead of the purely proportional one is

$$\lambda_i = 100\left(\frac{s_i - t_i}{s_i}\right) = 100\left(1 - \frac{t_i}{s_i}\right) \geq 100\left(1 - \frac{\frac{v_i}{N}(1-\alpha)S + 1}{\frac{v_i}{V}S - 1}\right).$$

Notice that, under the assumption that party i is not tiny, the last denominator is positive. Taking into account inequality (8.4), one gets

$$\lambda_i \geq 100\left(1 - \frac{2\frac{v_i}{V}(1-\alpha)S + 1}{\frac{v_i}{V}S - 1}\right) = 100\left(1 - \frac{2(1-\alpha)}{1 - \frac{V}{v_i S}} - \frac{1}{\frac{v_i}{V}S - 1}\right).$$

Finally, since $\omega_i = \frac{v_i}{V}$ one obtains

$$\lambda_i \geq 100\left(1 - \frac{2(1-\alpha)}{(1 - \frac{1}{\omega_i S})} - \frac{1}{(\omega_i S - 1)}\right). \qquad \square$$

For a fixed ω_i, the value of this threshold increases as α grows; it follows that as the system gets closer to the pure first-past-the-post, the minimal loss gets higher and higher, as expected. Moreover, for a fixed dosage α, the threshold is an increasing function of ω_i: hence, larger parties tend to bear larger percentage losses.

If in the above expression one sets $\alpha = 0.75$ and $S = 630$ (as in the current system for the Italian Chamber; see D'Alimonte and Chiaramonte, 1993), a party with 4% of the votes is doomed to bear a loss of at least 43% of the seats it would have received otherwise, and one with 7% of the votes must suffer a certain loss of about 47% of the seats! Even though some seat loss is predictable, the actual amount is stunning.

If the electoral system also includes an exclusion threshold, the loss is somewhat smaller. For example, if the threshold is 4% and if the parties below the threshold altogether get no more than 10% of the votes, then one sees that parties above the threshold lose at least 30.8% of their seats.

Finally, we notice that, when $S = 630$ and $\omega_i = 0.04$, the value of the above expression is negative whenever $\alpha < 0.53$. As a matter of fact, one could imagine some extreme (but unlikely) situation in which $t_i > s_i$ and thus $\lambda_i < 0$. This might happen when the mixed system is nearly proportional (α close to 0) and when in each district the best two candidates get about the same number of votes and the remaining candidates get a negligible number of votes. In situations of this kind, party i, due to the detraction rule, might actually *gain* votes in a mixed system.

8.3 Different Mixtures, Different Political Effects

The evaluation of the performance of an electoral system on the political structure of a country has been the main concern of political engineers since the work of Rae (1967). The careful study of many researchers has produced a detailed classification of the political consequences of the main seat-assignment methods, although the results in this field can only point out trends and broad indications, since a limited number of variables can actually be considered. Nevertheless, building hypothetical scenarios can help understand the effect a specific electoral system may produce in real life.

This is the kind of approach proposed in this section to analyze the performance of simple hybrid systems depending on the value of the parameter α, that is, the fraction of seats allocated by a plurality rule.

First, let us define a *simple hybrid system* as a seat allocation method that assigns $\alpha\%$ of the S seats with the first-past-the-post method and the remaining $(1-\alpha)\%$ with the d'Hondt method. Needless to say that the parameter α must be chosen so that αS (the number of single-member seats) and $(1-\alpha)S$ (the number of d'Hondt seats) be integers. The parameters of a simple mixed system are the following:

- a value for the dosage α, $0 < \alpha < 1$;
- the number of single-member districts, $k = \alpha S$;

- the number of ballots;
- the type of detraction rule.

Therefore, αS single-member districts must be drawn, subject to the criterion of population equality (fair apportionment), which is the standard procedure when plurality or majority rules are adopted. After assigning a seat in each of these districts, $(1 - \alpha)S$ seats must still be allocated. The simplest method is to consider a unique multimember district (at national or regional level) in the proportional allocation phase.

A typical electoral strategy carried out by groups of (presumably close) parties in a majoritarian context consists of the presentation of a common list of candidates. Suppose that if two or more parties make an electoral agreement in one district, they continue to present common lists in all of the single-member districts. This assumption is not restrictive, since many electoral laws state it as a necessary condition at least for large constituencies. Moreover, "independent" candidates will not be considered; that is, each candidate necessarily belongs to a well-defined list.

Assume that only one ballot is used both to assign single-member seats and to determine the number of votes to be adopted in the proportional allocation phase, according to the total detraction rule.

In conclusion, the specific, although not restrictive, assumptions considered in this model are the following:

- fair apportionment;
- the electoral agreement between parties must be the same in all districts;
- unique ballot;
- total detraction rule.

Electoral Strategies and Party Behavior

Different electoral systems induce political parties to behave in a different manner to optimize their electoral strategies. We will try to describe some of the most typical reactions a party will tend to have in both a majoritarian and a proportional context in order to better understand party behavior when a mixed system is adopted. Our basic assumption is the *rationality of the actors* in the political arena; i.e., every party and every elector will maximize a well-defined utility function. The utility for a party can be thought of as the number of seats it wins in the election. On the other hand, we know that voters want their preferred party to win the elections, but it is much more complex to formulate the utility of a voter. In the following paragraphs we want to show how the utility of a voter depends on many different features, among which is the electoral system itself.

At the beginning of the electoral campaign, presumably only part of the voters are already totally convinced of the choice they will make on the ballot.

The others are still trying to decide to which party, coalition, or candidate they will give their support. When a proportional electoral system is adopted, the various parties will try to keep separate and distinct as much as possible because they know that they are always guaranteed a part of the total seats, even if it may be very small. On the contrary, majoritarian systems will encourage the merging of parties that cannot rely on winning in any single-member district.

Suppose there are five different parties (called A, B, C, D, and E) ranging from the left to the right in the political arena (Figure 8.2).

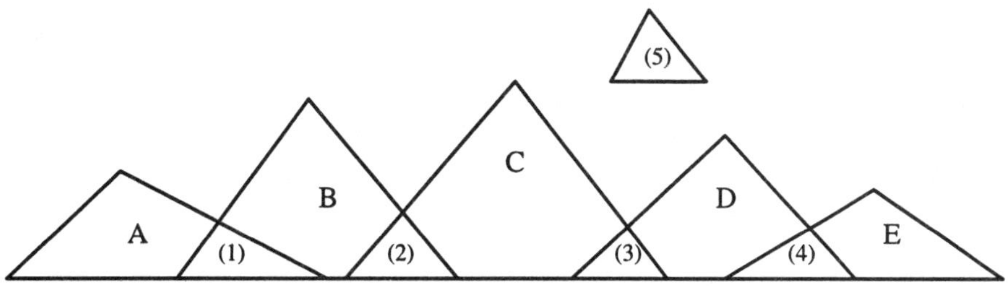

Fig. 8.2. *The voting body according to the parties voters are willing to support.*

The areas in capital letters represent the set of voters that have already decided to vote for party A, B, C, D, or E. The areas that are in the intersection between neighboring parties (denoted by numbers) represent the voters that generally float between two bordering political groups. Typically, at least in European countries, only a very small portion of voters are totally undecided on their political position. The totally undecided voters are represented in the figure by the area 5.

In general, the way the still undecided voters will react to any common electoral strategy (for example, the presentation of a common list of candidates) will differ depending on the kind of electoral system adopted.

Suppose parties B and C form a coalition. If the electoral system is proportional, the coalition (B+C) should expect to gain all the votes in area B, in area C, and in area 2. They probably will not get any vote from the citizens that belong to areas 1 and 3, which are located on their outer boundaries. In fact, although the electors in area 1 are more or less willing to vote for party B, they will feel that the coalition with party C is leading them too much to the right, while electors belonging to area 3 will think that party C has married political proposals that are too much to the left.

In such a case, parties B and C have to be extremely cautious: they will have to assess whether the coalition is profitable or not since the number of votes they would have obtained if they stayed separate could be larger than the number of votes they can expect by merging. On the other hand, if the first-past-the-post method is adopted, B and C will in many cases be forced to merge into a coalition if they want to survive. In fact, a coalition between B and C can expect to gain all votes in area B and C in addition to those in area 2 and most of those in areas 1 and 3 as well. Although they might prefer party A to the (B+C) coalition, voters in area 1 will feel the fact that D might win the single seat as a real threat and will eventually prefer to vote for (B+C) rather than leaving D such a chance. The same applies to the voters in area 3 who might prefer D to (B+C) but definitely do not want A to win. The idea is that in proportional systems coalitions are not convenient because everyone can get a portion of representation and nobody is clearly cut out. In majoritarian systems, coalitions are often necessary to contrast stronger opponents because only one seat is at stake. In hybrid systems, these two contrasting modes must be taken into account: since districts are single membered, parties will look for possible coalitions to prevent their opponents from winning the seat on one side; on the other, they will try to keep separate since coalitions could penalize them in the proportional seat assignment.

This brief overview tries to show that, in practice, while defining their electoral strategies, parties must always take into account two subordinate objectives. First, a party wants to maximize the number of seats it wins, and then it wants to maximize the number of seats the coalition it belongs to (if any) wins.

In a hybrid system, the use of a detraction rule makes everything even more blurred. Parties will want to form a coalition to support a common candidate, especially when such a candidate is closer to being a loser in the single-member district than a winner. This might sound like a paradox but it is due to two main reasons. On one hand, parties supporting a common defeated candidate will feel effective in fighting against the winning candidate since they force the winning candidate's party to remain with a very small amount of votes in the proportional seat assignment phase. In other words, parties try to minimize the probability that the party who wins the single-member seat also wins a large quota of proportional seats. On the other hand, if a party joins a coalition to support a common candidate who actually wins the seat in the single-member district, and the winning candidate does not belong to this specific party, the party will have to uphold a cost which might be excessively expensive. In fact, the detraction rule will make the party lose a quota of votes that could have otherwise been used to obtain a greater amount of proportional seats.

The Simulation Model

As mentioned above, in the electoral competition each party tries to maximize the number of seats it wins by a careful planning of its expenditures for the electoral campaign in the different districts.

Let b_i denote the available budget of party i; P_j the total number of voters in district j; v_{ij} the (unknown) number of voters for party i in district j; and y_{ij} the (unknown) expenditure of party i in district j. We assume that v_{ij} and y_{ij} are related by a linear model, as follows:

$$v_{ij} = \gamma_j y_{ij} + a_{ij}, \quad i = 1, 2, \ldots, n; j = 1, 2, \ldots, k, \tag{8.5}$$

where a_{ij} is the number of *faithful* voters for party i in district j (those electors who have a strong preference for party i and will give their vote to it in any case, without being influenced by the electoral campaign) and γ_j is the number of uncertain votes gained in district j per unit of expenditure.

The following constraints must hold:

$$\sum_{i=1}^{n} v_{ij} = P_j, \qquad j = 1, 2, \ldots, k, \tag{8.6}$$

$$\sum_{j=1}^{k} y_{ij} = b_i, \qquad i = 1, 2, \ldots, n, \tag{8.7}$$

$$v_{ij}, y_{ij} \geq 0, \qquad i = 1, 2, \ldots, n; \quad j = 1, 2, \ldots, k. \tag{8.8}$$

Taking into account (8.6), one could indifferently write the above constraints in terms of the variables v_{ij} or y_{ij}. In a preliminary set of experiments [Pennisi, 1996] the following—indeed quite strong—simplifying assumptions have been made on the linear model (8.5):

- $a_{ij} = 0, \quad i = 1, 2, \ldots, n; j = 1, 2, \ldots, k;$
- $\gamma_j = \gamma, \quad j = 1, 2, \ldots, k.$

Under these hypotheses, (8.7) implies that the total number of votes obtained by party i, that is, $v_i = \sum_{j=1}^{k} v_{ij}$, is proportional to the budget of such a party. In other words,

$$v_i = \gamma b_i, \quad i = 1, 2, \ldots, n.$$

In this case, scenarios on the budgets directly translate into scenarios for votes. The simulation exercise carried out is an attempt to understand the effect of a variation in the dosage α on the main indicators adopted to evaluate the performance on an electoral system, described in detail in Chapter 3. The motivation of the simulation is not so much that of providing a realistic overview of what happens when hybrid electoral systems are adopted but rather that of understanding their technical potential in terms of political indicators. This means we will analyze what happens in terms of representation, government stability, and so on even in the case where there are no constraints on a party's action, that is, in the case parties can decide where to place their supporters. If this were possible, a party would take into consideration two subordinate objectives, the only constraint being that the total number of votes each party can control has already been fixed at the national level. The two objectives can be defined as follows:

(1) dispose supporters so as to win in the maximum number of single-member districts without wasting votes;

(2) in those single-member districts where the party cannot win the seat, dispose voters so as to maximize the detraction the winning party will undergo if possible; otherwise, "abandon" the district.

In fact, a party will first try to win as many single-member districts as possible (given that the number of total votes it can count on is fixed), disposing of as many votes as possible in each district. However, to win a single-member district it is not necessary to control all the district votes but only to obtain one more vote than any other party. Additional votes would be wasted, so that each party in practice tries to *minimize* the number of votes it needs in each district to win the seat. Moreover, if a party is the second largest in a district, it cannot win the seat at stake, but it can penalize the winning party in the proportional assignment by maximizing the number of votes that must be detracted. Let v_{ij} be the (unknown) number of votes cast for party i in district j, v_i be the total number of votes cast for party i, and P_j be the total number of votes cast in district j. The constraints defined above, in this simplified model, are

$$\sum_{i=1}^{n} v_{ij} = P_j, \quad j = 1, 2, \ldots, k,$$

$$v_i \equiv \sum_{j=1}^{k} v_{ij} = \gamma b_i, \quad i = 1, 2, \ldots, n,$$

$$v_{ij} \geq 0, \quad i = 1, 2, \ldots, n; \; j = 1, 2, \ldots, k.$$

Moreover, consider the vector function

$$f\ (\boldsymbol{V}, \alpha) = (f_1(\boldsymbol{V}, \alpha), f_2\ (\boldsymbol{V}, \alpha), \ldots, f_n(\boldsymbol{V}, \alpha)),$$

which can be defined in order to determine the number of seats $f_i\ (\boldsymbol{V}, \alpha)$ assigned to party i on the basis of a simple mixed system with $\alpha\%$ single-member seats when the vote distribution is given by the vote-matrix $\boldsymbol{V}$. Function $f\ (\boldsymbol{V}, \alpha)$ cannot be written in an analytical form, but it can be described by a list of elementary operations that must be carried out to transform the votes into seats, as stated in the mixed system procedure described in Figure 6.6.

This problem can be viewed as a *noncooperative zero-sum game* with n players,[8] where the utility function corresponding to player i is determined by the function $f_i(V, \alpha)$. The simulation consists of determining the vote matrix $\boldsymbol{V}$ that is produced by taking into consideration the two subordinate objectives mentioned above, under different political scenarios characterized by the number n of parties in competition and the percentages of votes cast for the different parties (vector $\boldsymbol{w}$).

[8]Game theory is a precious tool in decision sciences. An *n-person game* is the formal representation of the alternative strategies *n players* can express in terms of their *gains* and *losses*. The players themselves need not necessarily be individuals, but they can represent compact groups of decision makers with a common goal. The game is *cooperative* if coalitions between the n players are allowed. Moreover, a game is *zero-sum* if the whole amount one player gains is equal to the sum of the amounts the other players lose. Brams (1975) is an excellent reference for game theory used in the political context.

Results

The simulation model was applied to nine different political scenarios. Each scenario is characterized by the number of parties and the total vote distribution (Table 8.3). These two variables determine the degree of fractionalization of the party system (where F is the Rae fractionalization indicator given in section 3.2). Only extreme cases are considered in the vote distribution: parties that have more or less the same number of votes on one side, and a few parties that can be considered big against many small parties on the other.

Table 8.3. *Political scenarios.*

Number of parties (n)	*% votes per party* (w)	*Electoral fractionalization* (F)
3	32%, 33%, 35%	0.666
3	20%, 35%, 45%	0.635
3	10%, 15%, 75%	0.405
4	23%, 24%, 26%, 27%	0.749
4	5%, 10%, 20%, 65%	0.525
5	15%, 18%, 20%, 22%, 25%	0.794
5	5%, 6%, 10%, 19%, 60%	0.588
6	14%, 15%, 16%, 17%, 18%, 20%	0.831
6	3%, 5%, 10%, 15%, 20%, 47%	0.703

This kind of experiment provides useful information on the political consequences we can expect a mixed system to produce. The main results of the experiment concern the relation between specific values of α and a set of performance indicators for disproportionality and government stability.

It was noticed that the seat distributions with larger variance correspond to lower percentages of single-member districts, whatever the number of parties and the degree of electoral fractionalization considered. This means that when α is small we should not expect to be able to make a reliable forecast of the number of seats each party will actually win, even when we know the number of votes it has received at national level. In any case, it always happens that disproportionality is larger when the number of single-member seats is low. The maximum degree of disproportionality occurs for $\alpha = 0.1$ or $\alpha = 0.2$, and it starts to decrease considerably only when $\alpha = 0.5$. This occurs whatever the initial political scenario, so as to say that hybrid systems where α is small generally affect the degree of representation. It has been observed that a small quota of single-member seats acts like a sort of majority bonus, in particular when in the political scenario the votes are not distributed equally among the parties.

Moreover, it was noticed that the degree of seat fractionalization is very difficult to modify significantly with respect to the initial vote fractionalization.

In general, it tends to increase when α increases, contrary to disrepresentation. Nevertheless, when the votes are more or less equally distributed among the parties in the initial scenario (when the electoral fractionalization is higher) a small improvement in terms of seat fractionalization is obtained only at the cost of a much larger loss in terms of representation.

The performance in terms of government stability has shown that the decrease in fractionalization generally results in the transfer of seats from smaller to larger parties. The aim of such a model is to analyze the potential of a hybrid formula to amplify or attenuate the tendency expressed by the vote distribution as a function of the parameter α. We are therefore interested in understanding if a hybrid method can produce a desired effect in a given political situation. To analyze this problem we have resorted to a totally unconstrained situation, leaving the competition between parties to resemble an "ideal" case in which they can dispose their votes where they prefer and they act in a rational manner. Obviously, such an assumption is unrealistic, but it provides interesting guidelines for the definition of electoral campaigning strategies and it stresses a first attempt in the theoretical analysis of hybrid systems.

Part III

Designing Electoral Districts

Gallia est omnia divisa in partes tres.
Julius Caesar, *De Bello Gallico*

The methods used to translate votes into seats have been discussed at length in the previous chapters. In this part we shall deal with another feature, the electoral district plan, which is of no minor importance in determining the final seat assignment. Designing the size and shape of electoral districts may sound like a merely technical question with no interesting consequences for the simple voter. Nevertheless, the way in which the country is divided and the number of seats assigned in each district play an important role in the overall outcome of the elections. In the following chapters we shall describe some of the traps or the ambiguous effects that are hidden in district plans and, furthermore, we will try to analyze the methods and techniques that can be used to prevent abuse and build "neutral" political districts. Electoral districts are the units within which the outcomes of the vote are translated into legislative seats. They are usually defined as territorial units, but sometimes (when the aim is the protection of ethnic or cultural minorities) they can be directly defined by population subgroups. The importance of a correct district map is due to the fact that its interaction with the rest of the electoral system, and with the electoral formula in particular, produces various effects on the seat assignment. These effects can be attributed to two different factors:

- the magnitude of the district determines the number of seats to be assigned within each district;
- the shape of the district determines the particular combination of votes to be considered in the seat assignment.

The effects on the electoral results of district magnitude—and in particular the distinction between single-member and multimember districts—are often underrated. Multimember districts can have as few as two seats each, but they may also correspond to over a hundred seats. Furthermore, districts that belong to the same district map may not have the same magnitude. District magnitude is important because it affects the degree of proportionality of the whole electoral system. In fact, if there is just one seat to assign in a district, no electoral formula can be expected to produce a proportional seat allocation within that district unless the number of competing parties is equal to one. The larger the number of available seats, the more flexible the combination of seats that can be assigned in order to represent the relative strength of the parties.

Empirical evidence for these statements has been sought by many scientists [Rae, 1967; Lanchester, 1981; Taagepera, 1986; Shugart 1989]. Many different measures or indexes of proportionality have been used to establish the degree of proportionality that should be expected from specific pairs of district magnitude and electoral formula (see also section 3.4). The general effect can be stated as follows: the smaller the magnitude of the district, the greater the seats loss of the smaller political parties. Sometimes, this loss is reinforced by the electoral body itself, which is psychologically discouraged from voting for smaller parties knowing that they have a smaller probability of winning a seat [Shugart, 1989]. In any case, the effect of district magnitude is strongly dependent on the kind of electoral formula adopted. Rae (1967) shows how disproportionality is inversely related to district magnitude (as the magnitude increases, deviation from proportionality decreases) and computes the linear correlation between these two variables on the basis of 115 real cases. The linear correlation is approximately equal to -0.37, and the relationship between deviation from proportionality and district magnitude is represented by the curve shown in Figure 1. Moreover, as the magnitude increases, proportionality increases, too, but at a decreasing rate.

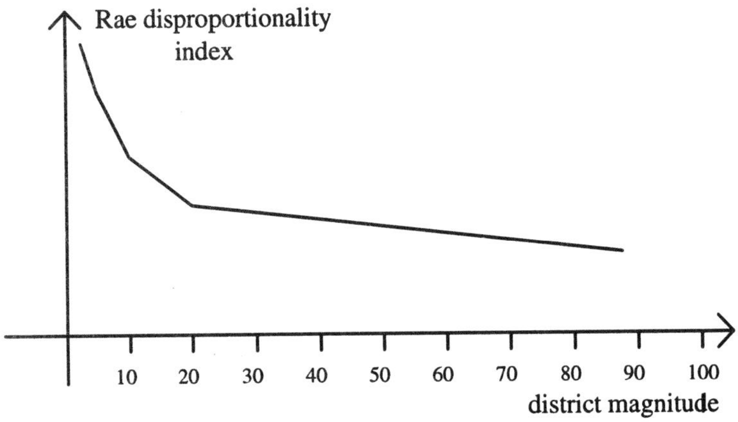

Fig. 1. *Disproportionality (Rae index) versus district magnitude.*

This means that the difference between districts with two and six seats is big in terms of the consequences on proportionality, but the same difference is smaller if we compare districts with 10 seats with districts with 20 seats, and it is definitely negligible between those with even more seats. In fact, the empirical evidence brought forth by Rae shows that up to magnitudes of about 20, small increases in magnitude are associated with declining but very substantial cuts in the average deviation between vote and seat shares. When district magnitudes are brought beyond 20, there seems to be a "plateau effect." In fact, magnitudes of 100 or 150 seats found in the Israeli and Dutch systems produce deviations that are just slightly below those generally found with district magnitudes between 10 and 20. Unfortunately, the data Rae examined did not present any case between 20 and 100 to strengthen his claims.

Part III begins with the analysis of the consequences of the shape of electoral districts and the relationship between electoral districts and the manipulation of electoral results through gerrymandering. Simple examples of gerrymandering and historical details on its occurrence are given in Chapter 9. In Chapter 10 we discuss the main criteria that must be considered in designing electoral districts, and we analyze the many indexes that are related to such criteria in Chapter 11. A good district map should be able to satisfy many criteria, such as population equality, compactness, conformity to administrative boundaries, etc. Some of these criteria are explicitly stated in constitutional laws, such as the British Reform Act (1832) or the United States Apportionment Act (1911). In any case, political districting must be considered as a multicriteria problem. In Chapter 12 we outline several old and new algorithms for drawing "neutral" political district plans. In particular, we suggest the use of local search techniques, such as simulated annealing, tabu search, and old bachelor acceptance, to find alternative district maps providing good values from a multicriteria point of view. In this context, the evaluation of trade-offs between criteria is crucial to compare and choose alternative district maps. Results obtained in a recent study on the design of the Italian electoral district map are used as an example.

Chapter 9

Traps in Electoral District Plans

When single-member districts are taken into account, a careful analysis of the shape of the districts may identify suspicious situations. If only one seat is at stake, the outcome may be radically changed by slightly modifying the boundaries of the district itself. History proves that there have been many attempts to manipulate the shape of electoral districts in favor of some political party or candidate. Although gerrymandering is an old technique, it is hard to kill, and it cannot be ignored when judging and drawing a district map.

9.1 Governor Gerry's Old Trick

Manipulation of political districts is popularly known as gerrymandering thanks to Governor Elbridge Gerry (1744–1814), who successfully managed to design the district map of the state of Massachussetts in order to guarantee his reelection. As the story goes, the future Governor of Massachussetts was defeated four times in a row between 1800 and 1803. Eventually, in 1810 and 1811, he was elected for the Federals and obtained the approval of a new district map to be used in the following elections. One of the districts had the shape of a salamander because it was designed specifically to include a sufficient number of his supporters in order to be sure he would win. Hence, the governor and the contraction between Gerry and salamander ("gerrymander") went down in history. Clearly, gerrymandering is easier to plan when the districts are single-membered since the risk of losing (even with a large number of votes) is very high. However, gerrymandering is not necessarily easy to detect. If you consider two adjacent districts and you know where your supporters are located, you can modify the district boundaries so as to include at least in one of the districts the minimum number of votes necessary to guarantee a seat. The idea is to manipulate the vote distribution by splitting up the adversary vote over many different districts and to concentrate a sufficient number of favorable votes in the same district. This can lead to drawing bizarre shapes and fantastic figures, which should be considered more appropriate to artists than to politicians.

To prevent abuse in political districting, several countries have decided to adopt "neutral" criteria such as population equality and compactness. Since the sixties, operations researchers have been interested in this tricky problem, and they have tried to attack it by making use of their favorite weapon—optimization.

9.2 Artificial and Historical Examples

The following examples are useful in understanding how district shape can affect the electoral result.

C	C	P	P	C
C	P	P	C	C
P	P	C	C	P
P	P	C	C	P
C	P	C	P	C
C	C	C	P	C
P	C	C	P	P
P	P	C	C	P
C	C	C	P	P

Fig. 9.1. *Territorial units and political preference.*

Dixon and Plischke (1950) consider an example where only two parties compete: party P and party C. The territory is divided into elementary units characterized by equal population and by the majority political preference, as shown in Figure 9.1.

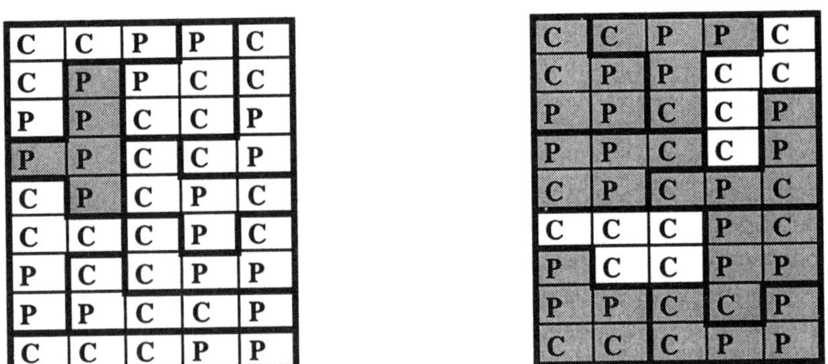

(a) (b)

Fig. 9.2. *In case* (a) *the winner is C; in case* (b) *the winner is P.*

In first-past-the-post systems, the way the territorial units are aggregated may have a drastic impact on the electoral outcome. Imagine having to draw a nine-district map where all districts have equal populations. The structure of the example is so simple that it allows us to find, even without the aid of a computer, many alternative maps that are perfectly uniform in terms of population. Two such maps are shown in Figure 9.2. But the map in case (a) makes party C win, awarding it eight districts against the single seat assigned to party P. The result is completely reversed in map (b), where P wins seven seats against two.

If this example is not convincing enough, consider the one illustrated by Giannuli (1992), where, given the vote distribution, four different district plans yield four completely different seat allocations. Suppose that a country is divided into 30 units, each unit being equal to the other in terms of population, and that there are two parties (A and B). Consider the table of votes given in Figure 9.3 (votes on the left are for party A; votes on the right are for B).

7-3	**4-6**	**9-1**	**9-1**	**9-1**
3-7	**4-6**	**4-6**	**4-6**	**6-4**
4-6	**6-4**	**4-6**	**6-4**	**4-6**
4-6	**4-6**	**4-6**	**4-6**	**3-7**
3-7	**4-6**	**4-6**	**7-3**	**3-7**
7-3	**5-5**	**6-4**	**5-5**	**7-3**

Fig. 9.3. *Votes for A and B.*

Summing up all the votes expressed in favor of one or the other party, we can see that party A has received 153 votes in total, while B has received 147.

Consider having to draw a 10 single-member district plan each containing 1/10 of the total population, that is, three territorial units each. Typically the districts must be connected, too, so only adjacent territorial units may be grouped together. Four different configurations, which all respect population equality and connectivity, are shown in Figure 9.4.

Figures 9.4(a) and (c) show a district plan in which party B wins against A and in which party A wins against B, respectively, with a difference of two seats in both cases. Figure 9.4 (b) shows a tie between the two, while in case (d) party B wins overwhelmingly.

The paradoxical case is (d) because 80% of the seats go to the party that has scored only 49% of the votes. On the contrary, case (a) seems more acceptable because it allocates more seats to the party that has more votes.

Many more examples can be cooked up (see, for example, Vickery, 1961). The situations they describe are artificial, but the problem they point out is extremely real and it cannot be ignored if one wants to guarantee neutral district plans.

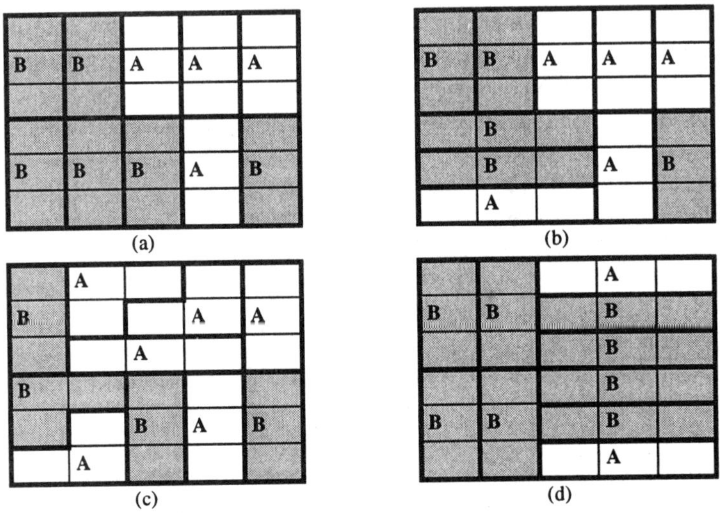

Fig. 9.4. (a) *A wins four seats and B wins six seats;* (b) *A and B both win five seats;* (c) *A wins six seats and B wins four seats;* (d) *A wins two seats and B wins eight seats.*

Real cases of gerrymandering keep showing up in recent history [Tentoni, 1991]. For example, it is believed that De Gaulle designed a partisan district plan for the elections of the French National Assembly in 1958 such that the votes he could rely on from his supporters would yield the most they could in terms of seats, thus penalizing the parties of the left which were his opponents. Since the electoral formula adopted was the double ballot, it is not easy to quantify exactly the importance of the manipulation hidden in the district configuration. On the other hand, it is possible to compare the vote distribution obtained after the first ballot and the final seat distribution. De Gaulle's party received 17.6% of the total votes at the first ballot, against 18.9% awarded to the Communist and 15.5% awarded to the Socialist. But after the second ballot, the outcome of the elections was 207 seats assigned to the Gaullists, only 10 to the Communist and 47 to the Socialist. Although the appeal to a second ballot always forces transfers of votes between parties, the radical difference between the first distribution of votes and the final distribution of seats leads to strong suspicions on the electoral districts.

We have already explained why gerrymandering is easier when the districts are single membered. Nevertheless, partisan districting is not excluded with multimember districts, too. In fact, this seems to have occurred in the 1981 elections for the Parliament of Malta. The electoral system adopted in Malta divides the island into 13 five-member districts and, in each district, the seats are assigned by a proportional formula with a single transferable vote. In the 1981 elections, the Labourist party managed to achieve 34 seats with 49.1% of the total vote, while the Nationalist party received only 31 seats even if it obtained 50.9% of the total vote.

Besides district manipulation, gerrymandering can arise without changing district boundaries but exploiting the demographic and geographic evolution of population. The main objective in gerrymandering is to alter the relation between total population and the share of population voting for a specific party. Migrations, immigration, and demographic evolution do this naturally. Neutral political districting must take this into account. A typical example of this phenomenon arises in the electoral history of the United Kingdom. In 1832, the changes in society due to industrialization and, in particular, the effects of migration from the countryside to towns made it necessary to rebalance the old electoral districts through the British Reform Act. The old districts were designed taking into account zones of the same area, causing a strong underrepresentation of the big industrial cities. Since then, population equality has become a universally accepted criterion for the planning of electoral districts.

Chapter 10

Criteria for Political Districting

In this section we survey the most important and common political districting criteria. The definition and formalization of appropriate criteria for the districting problem is in itself a difficult task, at least as difficult as the selection of a smaller set of criteria to use while designing the district map. It must be always kept in mind that the choice of the districting criteria affects the final solution.

Some constitutional laws explicitly state which criteria must be adopted in the districting procedure. Nevertheless, there are many alternatives to choose from and a rich literature on this topic has flourished in the past 35 years, since political districting has started to be considered a mathematical problem.

The most important characteristic of the criteria we will examine below is that they can be considered *neutral* or *objective*. A criterion can reasonably be considered "neutral" if it does not depend on the past electoral results and it cannot be used to steer the future ones. Neutrality is important in order to provide a clear, transparent, and nonpartisan solution to the districting problem, although nonneutral criteria may be taken into account for special purposes, such as the protection of cultural, ethnic, and racial minorities.

10.1 Integrity, Contiguity, and Absence of Holes

Most of the criteria considered in political districting are strictly connected to the geographic features of the district, such as its shape and its size, and to the structure of the territory, such as natural barriers and road networks. In a mathematical model a criterion can appear as a goal to reach or it can be formulated as a constraint. In particular, elementary criteria that are nearly always adopted, integrity, contiguity, and absence of holes, are generally formalized as constraints in the mathematical model.

Integrity requires that a territorial unit be not split between two districts. When a graph model is adopted to represent the territory [Nygreen, 1988;

Arcese, Battista, Biasi, Lucertini, and Simeone, 1992], the natural association of a node with an elementary territorial unit will guarantee integrity to be automatically satisfied.

A district map is *contiguous* if every district is made of a single land portion such that it is possible to walk from every point in the district to any other point in the same district without having to leave the district itself. Figure 10.1 shows a contiguous district map, while in Figure 10.2 contiguity is not satisfied since district C is made of two portions of land (C_1 and C_2). In fact, given a point in C_1 and another point in C_2, these two can be connected only by passing through district A or B, or both A and B.

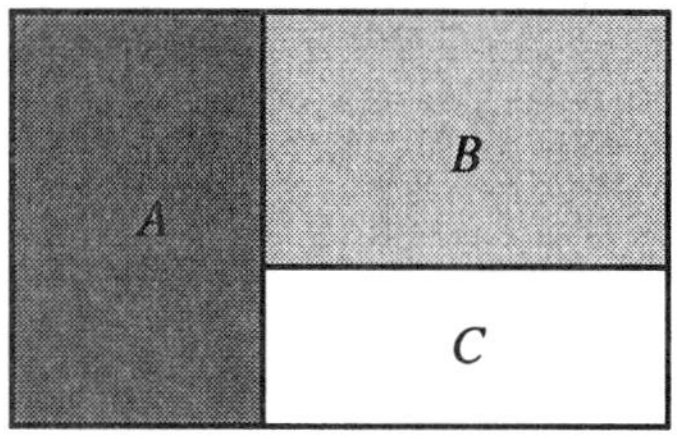

Fig. 10.1. *The district map is contiguous.*

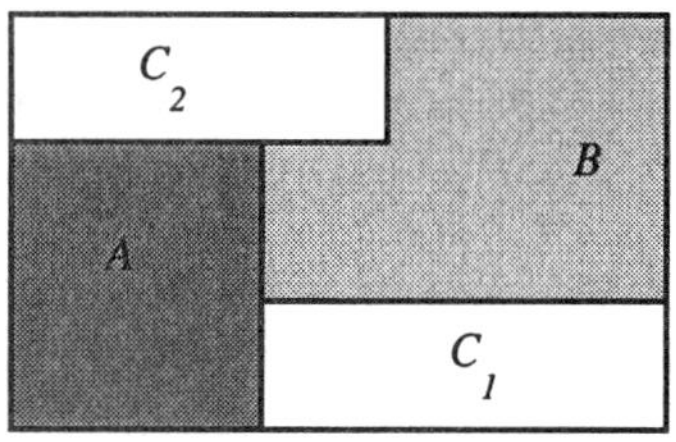

Fig. 10.2. *The district map is not contiguous.*

Another important criterion is the *absence of holes.* This criterion is satisfied when the following property holds: if one draws any closed curve C in a given district, all points within the inner domain of C still belong to the district. In particular, a distinction can be made between the concept of *hole* and the more specific idea of *enclave.* Let us define an *enclave* as a district that is fully surrounded by another district; more formally, a contiguous district is an *enclave* if its boundary is completely contained in the boundary of exactly one different district. Figure 10.3 shows examples of different types of holes in a district map.

Contiguity and absence of holes are mainly intended to prevent the geographic features of a territory to be exploited in order to draw a district map that favors a specific political party. District maps that abuse the territory, putting together pieces of land that best fit some political bias, ought to be forbidden. The same point can be made for the compactness criterion, which will be described later. Actually, contiguity and absence of holes are strictly related

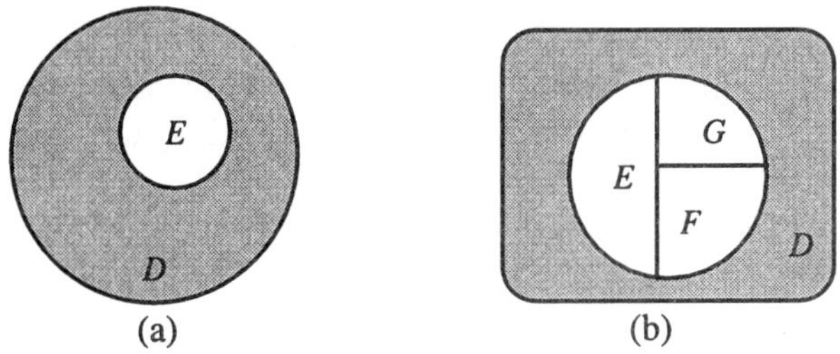

Fig. 10.3. (a) *D and E form a contiguous map, but E is an* enclave *and it forms a hole in D;* (b) *D, E, F, and G form a contiguous district map, but E, F, and G form a hole in D.*

to compactness since the existence of either holes or noncontiguous districts often corresponds to bad values for most indicators of compactness.

10.2 Population Equality, Fair Apportionment, Compactness, Administrative Conformity, and Socioeconomic Homogeneity

Population equality, fair apportionment, compactness, conformity to administrative boundaries, and socioeconomic homogeneity are often formalized as objectives in districting models. All these criteria embody a different idea of fairness that seems desirable for a district map. Nevertheless, some of them appear to be more important than others. As discussed in the following paragraphs, in a single-member district map population equality is undoubtedly the main criterion because it requires the districts to be balanced with respect to the population. Although very difficult to handle, compactness is one of the most important criteria because it has a key role in the prevention of gerrymandering.

Population Equality and Fair Apportionment

Population equality is the best assessed criterion in political districting procedures. It is widely accepted by all the authors who studied this kind of problem since its origins [Hess, Weaver, Siegfelatt, Whelan, and Zitlau, 1965; Kaiser, 1966; Garfinkel and Nemhauser, 1970]. Population equality requires the population size of each district to be equal to the average population $\bar{p}$, i.e., the total population divided by the total number of districts in the map. In a single-member district map the number of districts is equal to the total number of seats that have to be allocated, so, if population equality is fully satisfied, all districts will have exactly the same population. This means that, independent of other variables that interact in the electoral procedure (such as party size, coalitions, etc.), each individual vote (each voter) contributes in the same proportion to the allocation of the seat in the district. Let P_j be the number of electors in district j; then each vote should weigh exactly $\frac{1}{P_j}$. Moreover, if we consider the total number of seats S and the total population P, the ratio $\Gamma = \frac{P}{S}$ gives a rough evaluation of the population per seat and, ideally, each seat should correspond to Γ votes. Therefore, when a single-member district map satisfies population equality perfectly ($P_j = \bar{p}\ \forall j$) the ratio number of voters to number of seats in each district perfectly coincides with Γ ($P_j/S_j = \bar{p}/1 = P/S$).

The same can be required in multimember district maps. In fact, fair apportionment can be considered just as an extension of population equality. A district map satisfies the fair apportionment criterion if each district population size is a multiple of Γ, i.e., if S_j is the number of seats to assign in district j, then $P_j = S_j\Gamma$. Notice that, when $S_j = 1$ we have population equality for the single-member district case.

Now, if $P_j = S_j\Gamma\ \forall j$, then $\frac{P_j}{S_j} = \Gamma$; i.e., in every district a seat corresponds to exactly Γ voters as happens at the national level.

Population equality and fair apportionment embody the principle of one-man-one-vote in the single- and multimember district case, respectively. Clearly, population equality does not fit the multimember case because different amounts of seats to be allocated make a difference in the political weight of the individual vote. In particular, given a fixed population size, the ratio between total population and the number of available seats is higher the smaller the number of seats to be assigned in a district. Thus, it is important to guarantee that population and number of seats be proportional with the same coefficient of proportionality Γ in each district.

Compactness

Compactness is a very intuitive concept, but, unfortunately, a rigorous definition of compactness does not exist. A district (or in general a figure) can be considered compact if it spans a round region, without straggling or rambling, but it is hard to say what compactness actually means.

Deviation from compactness has been classified [Taylor, 1973] according to the shape of the district in the following four categories: elongation, indentation, separation, and puncturedness. A figure is *elongated* if it is narrow and long; it is *indented* if the boundary is irregular, with many protrusions and recesses; it is *separated* if it consists of several separate pieces; it is *punctured* if it has holes. Mathematicians would perhaps call a figure indented if it is nonconvex, separated if it is not connected, and punctured if its genus is positive. Notice that Taylor's definition of compactness is very strong, since it implies both contiguity (nonseparation) and absence of holes (nonpuncturedness).

Caution should be used when dealing with this criterion because every attempt made to define a correct measure of compactness turns out to be strongly related to one of the above categories, to the point that each indicator is able to recognize only some types of noncompactness. In section 11.2 we will focus on the construction of compactness indexes and we will show several tricky examples.

Most authors deem compactness necessary in the districting procedure even if no one agrees on which measure should be used. Some suggest dropping a rigorous definition of compactness because the choice of a wrong indicator can be worse than choosing to definitely ignore it. After drawing the district map, its degree of compactness should be evaluated by common sense, case by case, in order to modify those districts that clearly appear to be *dangerously* noncompact. In our opinion, this kind of approach leaves too much room for personal (and possibly erroneous) interpretations of compactness.

An idea could be to take more than one type of compactness into account at the same time. For example, we can choose an index based on perimeter to measure indentation, one based on area to measure elongation, and a third to measure separation. This implies a careful measurement of compactness at the expense of a very complicated mathematical model which could turn out to be unmanageable if there are also other objectives. This is why, in the end, the most practical thing to do might be to measure compactness, as every other criterion, with a single index.

Conformity to Administrative Boundaries

A district map satisfies *conformity to administrative boundaries* if each district does not cross the boundaries of given administrative areas, such regions, provinces, health-care districts, etc. For example, a district fully satisfies conformity to *regional* boundaries if all territorial units lie in the same region. Conformity holds in both cases (a) and (b) shown in Figure 10.4, but not in case (c) where the district intersects an administrative area and neither the district nor the region is totally contained in the other.

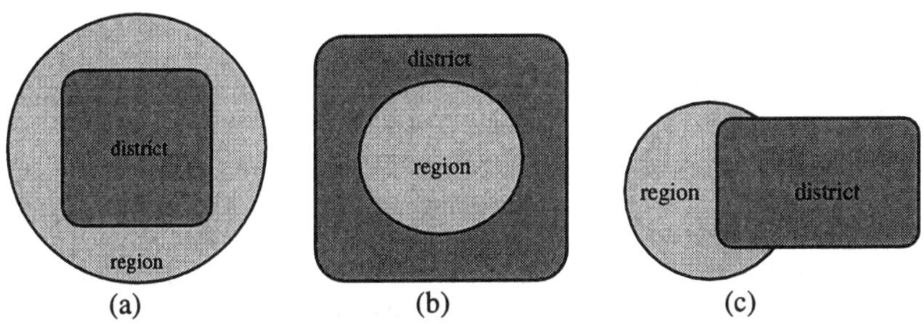

Fig. 10.4. (a) *The district is totally included in the region;* (b) *the district contains the whole region;* (c) *the district and the region overlap.*

When the whole district map is considered, perfect conformity to administrative boundaries is possible only in exceptional cases. In some cases, several administrative boundaries are considered at the same time. Respecting such boundaries can be useful to simplify electoral procedures such as the identification of the electoral body and other organization issues.

The appeal of administrative boundaries is that, in many countries, they provide an initial map for the electoral districting procedure. In fact, administrative boundaries are often defined on the basis of population and geographic features, as well as on history. However, this is usually not enough to guarantee the satisfaction of other important criteria, too, such as population equality and compactness.

A suitable indicator to measure the degree of conformity to administrative boundaries should be able to recognize *good* and *bad* district maps on the basis of situations such as that shown in Figure 10.4(c). In section 11.3 some interesting indicators will be analyzed and discussed in detail.

Socioeconomic Homogeneity and Heterogeneity

Whether to adopt a socioeconomic criterion or not in a districting procedure is a very controversial issue. Some authors suggest using such criterion by requiring the districts to be as homogeneous as possible according to a specific set of fixed socioeconomic variables. Others are rather inclined to believe that socioeconomic heterogeneity must be required. Anyway, whatever the meaning ascribed to this criterion, it obviously cannot be defined neutral, because the socioeconomic and cultural characteristics of the electors are necessarily correlated with their vote.

This criterion is able to influence the electoral system and it can be the cause of unexpected electoral results. Suppose that an election is to take place on planet Tschai (the well-known creation of science fiction writer Jack Vance), where four different ethnic groups (Pnume, Dirdir, Chash, and Wankh), with very different cultural characteristics, coexist, although in different parts of the planet. Suppose the first-past-the-post system is adopted; therefore, single-member districts must be designed. Population equality must be considered in order to respect the one-man-one-vote principle. If socioeconomic homogeneity is adopted as well, each district will not only be of the same size but people belonging to the same cultural group will be gathered in the same district. Since it can be assumed that in such a situation electors of the same subgroup vote for the same party (the one that best represents them), then, on the basis of population equality, the number of seats obtained by each party will turn out to be proportional to the number of citizens of each different ethnic group. Therefore, the electoral result provided by a majority formula will be the same as the one a pure proportional method would have provided in a unique multimember district.

On the other hand, if districts are required to be diversified with respect to a set of socioeconomic variables (*socioeconomic heterogeneity*), neutrality may still not be respected. In fact, if a specific socioeconomic composition is required in each district, all seats will turn out to be assigned to the party that represents the group with the largest fixed percentage of population in the single-member district.

From this brief discussion it is clear that socioeconomic homogeneity and heterogeneity both strongly interfere with the electoral system and affect the electoral results. Nevertheless, in some cases such criterion is essential. An interesting example is given in Bourjolly, Laporte, and Rousseau (1981) who designed the district map for elections in the Ile de Montréal, where the adoption of socioeconomic homogeneity is justified by the purpose of ensuring an adequate representation to ethnic minorities.

Chapter 11

Indicators for Political Districting

In the previous section, we discussed why more than one criterion should be adopted in a districting procedure. We reviewed the most important criteria to understand their importance in assessing the fairness of the final district map. This section deals with a more complex problem, that is, the definition of correct indicators to measure the criteria described above. As we have pointed out, some criteria involve several different features and cannot be defined in a unique manner. This makes the definition of suitable indexes more difficult, but, in any case, a good formalization is always very important for a satisfactory result, whatever the districting procedure adopted.

11.1 Population Equality

The most popular indexes of population equality are global measures of the distance between the populations of the districts and the mean district population $\bar{p}$. The most simple index measuring the lack of equality between district populations is defined as

$$\frac{\sum_{j=1}^{k} \left| P_j - \bar{p} \right|}{k}, \tag{11.1}$$

where P_j is the population of district j, and k is the total number of districts. Other indexes can be built simply by replacing the L_1-norm by other norms. For example, L_2 gives the population variance:

$$\frac{\sum_{j=1}^{k} (P_j - \bar{p})^2}{k}. \tag{11.2}$$

This index has the advantage of penalizing large deviations from $\bar{p}$ more than small ones.

Unfortunately, the range of these measures depends on the size of the total population, so relative measures with values in the [0,1] interval are usually preferred. An interesting index that was recently defined by Arcese, Battista, Biasi, Lucertini, and Simeone (1992) considers the sum of the absolute values of the deviation from the average population divided by the maximum value, as follows:

$$\frac{\sum_{j=1}^{k}\left|P_j - \bar{p}\right|}{2(k-1)\bar{p}}. \tag{11.3}$$

In fact, the maximum lack of population equality occurs when the whole population is assigned to one district, leaving the others totally empty. In this case,

$$\sum_{j=1}^{k}\left|P_j - \bar{p}\right| = \left|P - \bar{p}\right| + (k-1)\left|0 - \bar{p}\right| = 2(k-1)\bar{p}, \tag{11.4}$$

since the total population can be also written as $P = \bar{p}k$. In practice, however, index (11.1) almost always takes values in [0,1] and, in our opinion, is to be preferred to (11.3), also in view of its simple interpretation as mean absolute deviation of district populations from the ideal population $\bar{p}$.

A very different index is given by the *inverse coefficient of variation* (ICV) [Schubert and Press, 1964], which is based on the standard deviation of a statistical variable with mean equal to 1 (*the coefficient of variation*). In our case, the variable is the ratio $\frac{P_j}{\bar{p}}$; thus the coefficient of variation is

$$c = \sqrt{\frac{\sum_{j=1}^{k}\left(\frac{P_j}{\bar{p}} - 1\right)^2}{k}}, \tag{11.5}$$

and ICV is defined as follows:

$$ICV = \frac{1}{1+c}. \tag{11.6}$$

While the indexes mentioned up to now measure the lack of population equality, this index yields higher values the more the population is equally distributed. It is always positive and it varies in $[1/(1+\sqrt{k-1}), 1]$. It reaches its minimum value when c is maximum ($c = \sqrt{k-1}$), while it increases up to a maximum value equal to 1 when c decreases toward zero (perfect population equality). The range of values for ICV thus strongly depends on k and this may be considered as an undesirable feature.

Kaiser (1966) suggests an index based on the geometric average of the ratios $\frac{P_j}{\bar{p}}, j = 1, 2, \ldots, k$. The simple geometric mean $g = \sqrt[k]{\prod_{j=1}^{k} \frac{P_j}{\bar{p}}}$ is a measure of uniformity which grows as population equality increases in the district map. Although g takes values in the [0,1] interval, values quite close to 1 are reached even under substantial malapportionment. For this reason Kaiser suggests an index that is functionally related to the geometric mean but that is also able to recognize unfair district maps. The index is defined as follows:

$$1 - \sqrt{1 - g^2}, \tag{11.7}$$

and its range is the same as g.

11.2 The Many Facets of Compactness

In the last 25 years the literature about compactness indexes has been so rich that many authors have dealt with the analysis and the comparison of different measures to understand which should be considered the best. In a recent paper [Horn, Hampton, and Vandenberg, 1993] over 34 compactness indicators are listed! Unfortunately, as mentioned in section 10.2, compactness is not easy to define and there still is no general agreement on the index to adopt.

The compactness of a district depends on its area, the distances between territorial units and the district center, the perimeter, the geometrical shape, its length and its width, the district population, and so on. On the basis of these features, different compactness indexes can be classified [Niemi, Grofman, Carlucci, and Hofeller, 1990] as shown in Table 11.1.

Table 11.1. *Classification of compactness indexes.*

Dispersion measures	length versus width; district area compared with the area of "canonical" compact figures (for example, the circle); moment of inertia;
Perimeter-based measures	perimeter; perimeter compared with area;
Population measures	district population compared with the population of the smallest compact figure (for example, a circle) that contains the whole district; population moment of inertia.

Examples of the most important and widely cited indexes are described in the following paragraphs. On the basis of the work carried out by Young

(1988), our aim is to show that, although they are all reasonable measures of compactness, there always exist special cases in which they do not work.

Reock Index

The Reock measure [Reock, 1961] is defined as the ratio between the district area and the area of the circle which circumscribes it. The index takes on values in [0,1] and higher values correspond to higher degrees of compactness. Unfortunately, the Reock index might be close to 1 even if compactness is completely absent. A striking example is a district shaped like a *serpent* (Figure 11.1(a)): in this case the district area and the area of the circle nearly coincide, although one would probably never regard such a district as being compact. Moreover, the Reock index fails to recognize noncompactness due to holes in the district (Figure 11.1(b)).

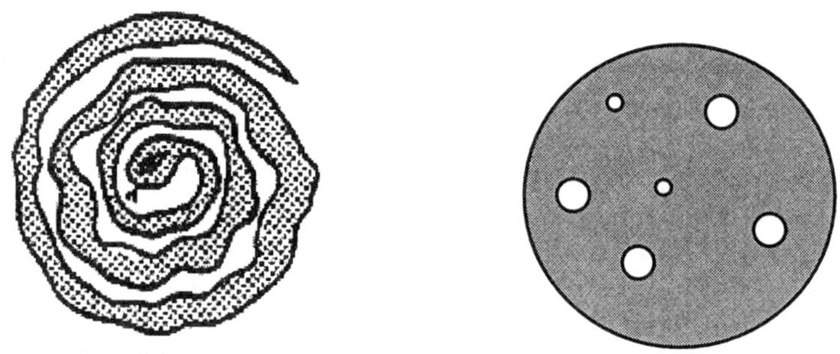

(a) (b)

Fig. 11.1. (a) *Serpent (after Young,* 1988*);* (b) *Gruyère cheese.*

Haggett Index

This index is based on the ratio of the radius of the largest circle inscribed in the district to the radius of the smallest circumscribing circle. This kind of index will easily detect noncompactness of serpentlike districts, since the radius of the inscribed circle will be very small and concentrated in the serpent head. However, it is not immune to other paradoxes. For example, a *pear*like district will be considered very compact because the ratio will be close to one (Figure 11.2).

Schwartzberg Index

The idea embodied in the Schwartzberg index [Schwartzberg, 1966] is, again, that the ideal shape for a district is a circle. It is well known that a circle has minimum perimeter among all plane figures with a given area. Thus, for any given district, the ratio between the perimeter of the district and the length of a circle having the same area may be taken as an index of noncompactness. Since calculating the district perimeter might be technically difficult, Schwartzberg

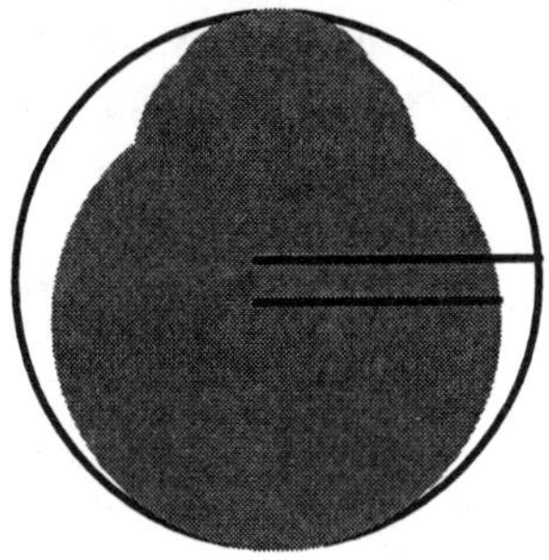

Fig. 11.2. *Pear district.*

suggests replacing the district boundary B by a polygonal approximation as follows: identify all those points along B where three or more territorial units (within the district or not) meet (in Schwartzberg terminology these points are called *trijunctions*); then join each such point with the next one along B, in clockwise order, by a straight line segment. The perimeter of the resulting (possibly nonconvex) polygon is the *adjusted* (or *gross*) perimeter of the district. The Schwartzberg index is given by the ratio between the adjusted perimeter and the length of the circle with the same area as the district. Nevertheless, such an index gives too much importance to the perimeter of the district and not enough to its shape. In fact, consider the case in which territorial units are small squares. A *puzzle-piece* district (which is therefore nearly circular) might turn out to be less compact than a *dog-leg* district just because its boundary is very irregular (Figure 11.3(a) and (b)).

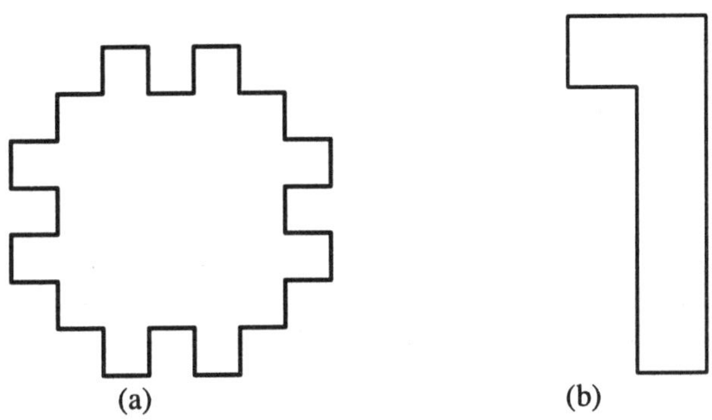

Fig. 11.3. (a) *Puzzle piece;* (b) *dog leg (after Young,* 1988*).*

Length-Width Index

The length-width index [Harris, 1964; Papayanopoulos, 1973] is the ratio of length to width of the rectangle circumscribing the district. Consider a *salamander* district (Figure 11.4): in this case the rectangle which circumscribes the district is a square, so the length-width index is equal to 1 and shows a situation of perfect compactness, but of course the salamander is strongly noncompact because of its high degree of nonconvexity.

Fig. 11.4. *Salamander (after Young,* 1988*).*

The evaluation of compactness must be made not only on the basis of the shape of single districts but also on the district map as a whole. In this case, other difficulties may arise. For example, consider a square territory: it seems impossible to divide the square into two convex districts such that both of them are perfectly compact with respect to the length-width index. Certainly it is absurd that the map in Figure 11.5(a), which is clearly compact, turns out to be worse than the one in Figure 11.5(b), which would be declared *perfectly* compact according to the length-width index.

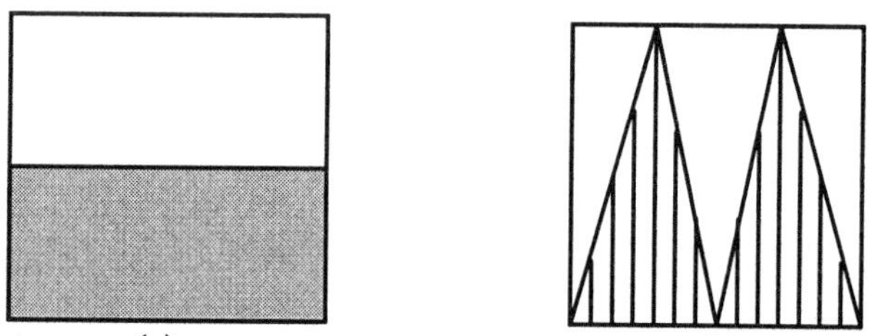

(a) (b)

Fig. 11.5. (a) *Equal size convex districts;* (b) *sawtooth (after Young,* 1988*).*

Taylor's Indentation Index

A very different point of view is given by Taylor's index, which focuses on indented noncompact figures [Taylor, 1973]. It is based on the comparison between "reflexive" and "nonreflexive" angles of the adjusted perimeter of the district (defined just as in Schwartzberg), where reflexive angles are those that bend away from the district. If R is the number of reflexive angles and NR the number of nonreflexive angles, then the index is

$$\frac{NR - R}{NR + R}. \tag{11.8}$$

However, this index performs very badly with noncompact regular figures such as a rectangular narrow strip.

Arcese, Battista, Biasi, Lucertini, and Simeone Compactness Index

This index [Arcese, Battista, Biasi, Lucertini, and Simeone, 1992] is inspired by the Reock index since it is based on the comparison between the shape of the district and a round figure. Consider a district D, its center c, and the farthest territorial unit u within the district (Figure 11.6). Draw the circle with center c and radius given by the distance between c and u. It is preferable to use road distances, rather than Euclidean (or "as the crow flies") ones. Such a circle obviously includes all of the territorial units that belong to district D, but it may also contain territorial units belonging to other districts. Now compute the percentage of units that are in the circle but not in the district. This percentage is a measure of (non-) compactness of district D. If D is round, its value is close to zero, while it is close to one if D is *banana-* or *octopus*-shaped.

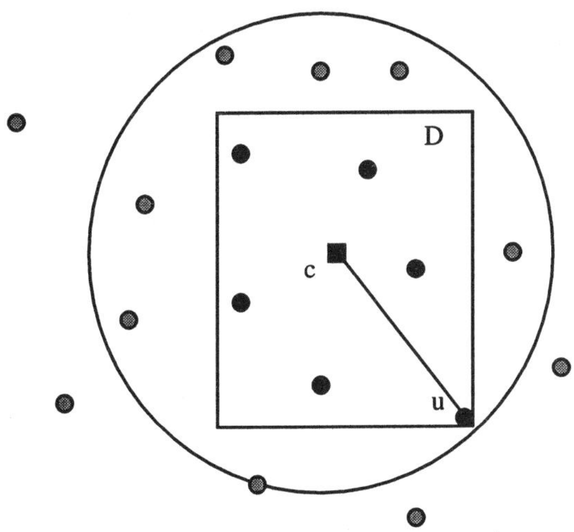

Fig. 11.6. *Arcese, Battista, Biasi, Lucertini, and Simeone compactness index.*

This index can be refined so as to evaluate compactness also with respect to population since each territorial unit can be weighted by its population. The weighted center and the percentage of population in the circle but not in the district are considered.

In this case, both the geographical location of the units and the distribution of their population are taken into account. Let P_i^u be the total population of the units within the circle; then the compactness index is the following:

$$\varphi = \sum_{j=1}^{k} \frac{P_j^u - P_j}{P_j^u}. \tag{11.9}$$

A very similar index [Grofman and Hofeller, 1990] is the ratio of the district population to the population in the minimum circumscribing circle. Notice that this index is equal to the previous one when the weighted center of the district corresponds to the center of the circumscribing circle.

Moment of Inertia

Hess, Weaver, Siegfelatt, Whelan, and Zitlau (1965) suggest measuring compactness by the moment of inertia. Let c be a point in district D. The moment of inertia of district D with respect to c is the weighted sum of the squared distances of all territorial units u in D from c. The weight of each distance is given by the population of the corresponding territorial unit. The moment of inertia is minimized by setting c equal to the *center of gravity* (g) of the district. To compare the compactness of different districts, the moment of inertia is always calculated with respect to g. The smaller the moment of inertia about g, the more compact the district.

If n_j is the number of units in district j and $d_i^{g(j)}$ is the distance between unit i in district j and the center of gravity of the same district, then the moment of inertia in district j is

$$\sum_{i=1}^{n_j} p_i (d_i^{g(j)})^2. \tag{11.10}$$

This index is rather different from those examined above because it does not use the comparison between the district and a "compact" geometric figure to determine the degree of compactness of the district. Moreover, it has the advantage of including population as a basic parameter for the evaluation. In this case, a district is considered compact if the territorial units are gathered around its center of gravity but also if the population distribution is concentrated in the units that are the closest to the center. Unfortunately, the moment of inertia is unable to recognize noncontiguous situations (again, consider the serpent in Figure 11.1), because it strongly depends on the territorial extension of the district.

Consider the typical example of rural and urban districts. Usually, population density in urban districts is larger than in rural ones, which implies that a rural district will generally cover a larger area with respect to an urban district with the same population. This means that two districts with the same population and shape could turn out to have different moments of inertia, simply because the distances between units and the center of gravity are very different.

In this case, it could be useful to adopt a scale-invariant index, which cancels the distortion caused by the dimension of the district. Let d_{max}^j be the distance between the center of gravity of district j and the territorial unit u in district j which is the farthest from $g(j)$. Divide each distance by d_{max}^j and call this ratio $\bar{d}_i^{g(j)}$. Then instead of (11.10) consider

$$\sum_{i=1}^{n_j} p_i (\bar{d}_i^{g(j)})^2. \tag{11.11}$$

At this point a global compactness index can be defined as the average value of the compactness measures referring to the single districts of the map. If k is the total number of districts, then

$$\frac{1}{k}\sum_{j=1}^{k}\left(\sum_{i=1}^{n_j} p_i \left(\bar{d}_i^{g(j)}\right)^2\right) \tag{11.12}$$

is the *average moment of inertia*.

The examples we have stressed up to now prove how difficult it is to say that a specific index is better than the others. Young (1988) draws the conclusion that compactness should not be used as a criterion, period. We feel that, without being so extreme, a reasonable attitude can be taken by recommending (as Taylor (1973) does, among others) to consider at the same time a small number of different indexes, each focusing on a particular feature of compactness, and to define compact only those districts that perform well for *all* of the measures taken into consideration. An example is provided by Garfinkel and Nemhauser (1970), who consider compactness as one of the constraints of the political districting model, but they formulate it by adopting two different compactness measures: both the distances between units of the same district and the district area are considered.

11.3 Finding Appropriate Indicators of Conformity to Administrative Boundaries

Conformity to administrative boundaries requires district maps to take into account administrative boundaries that already exist, such as regions, provinces, counties, and health-care districts. If only one type of administrative boundary is considered (say, the regions), the ratio between the number of regions that some districts overlap, as in Figure 10.4(c), and the total number of regions in the country is a simple index of conformity to administrative boundaries. The average value of these ratios, over all the different types of administrative areas under consideration, provides an index for the whole district map. Unfortunately, such an index is not very sensitive to small modifications of the district map.

For example, the above index is unable to discriminate between case (a) and case (b) in Figure 11.7. In both cases district 1 and district 2 are violating the boundaries of one region. Therefore, they both will be counted in the numerator of the index. However, case (b) should be considered better than (a) if we are interested in how much the administrative boundaries are violated.

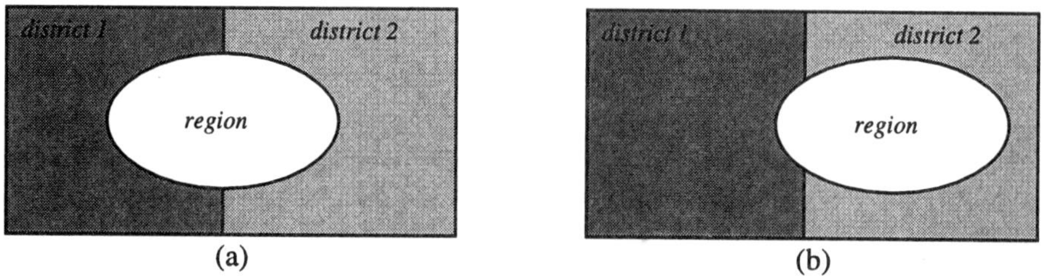

Fig. 11.7. (a) *Half of the region is in district* 1 *and the other half is in district* 2; (b) *the region is almost completely contained in district* 2.

A more sensitive index is based on the computation of the number of units in a district that belongs to the same administrative area. More precisely, given a territory and its district map, consider just one type, say, h, of administrative area at a time (for example, h may indicate the provinces). Let L_h be the total number of different administrative areas of type h (following our example, it can be assumed that the territory covers L_h different provinces). For a single district D_j in the map, let x_{jl} be the number of units in district j that are in the lth area of type h. Of course we have

$$\sum_{l=1}^{L_h} x_{jl} = c_j, \tag{11.13}$$

where $c_j = |D_j|$. For district D_j an index of conformity with respect to the area of type h can be defined as follows:

$$C(D_j, h) = \frac{\sum_{l=1}^{L_h} x_{jl}^2}{c_j^2}. \tag{11.14}$$

Notice that if D_j contains only one of the type h areas (and this is the best situation one can hope for), then $C(D_j, h) = \frac{c_j^2}{c_j^2} = 1$; on the other hand, the worst situation takes place when the c_j units are equally distributed among the L_h areas of type h, that is, $x_{jl} = \frac{c_j}{L_h}$ for each possible l. In this case, we have

$$C(D_j, h) = \frac{\sum_{l=1}^{L_h} \frac{c_j^2}{L_h^2}}{c_j^2} = \frac{1}{L_h}.$$

Thus

$$C(D_j, h) \in \left[\frac{1}{L_h}, 1\right].$$

A global administrative conformity index can be defined as the average of $C(D_j, h)$ over all districts D_j and all types of administrative areas considered.

Chapter 12

Optimization Models for Electoral Districting

This chapter focuses on optimization models and algorithms for electoral districting. In the last 35 years the literature on political districting has flourished, providing a variety of techniques that try to solve the problem efficiently. The many algorithms proposed consider more than one criterion, but all the earliest ones adopt a single-objective approach. Only more recent works attempt the way of multiobjective optimization. In our opinion, there are at least three main criteria that should be taken into account in order to produce a "good" district map. They can be viewed as constraints or as objective functions, but we prefer the second approach to avoid restricting the area of solutions by automatically cutting out some potentially good solutions by infeasibility. A multiobjective approach allows more flexibility: it offers the possibility to choose on the basis of real values of the criteria, while an a priori definition of thresholds forces the search toward a specific region of the feasible set.

After describing several districting algorithms and, in particular, the main multicriteria models found in the literature, we discuss *Pareto optimality* and define *local Pareto optimality*. Finally, we analyze trade-offs between criteria when more than one solution is available.

12.1 How to Handle Conflicting Criteria

An inspection of the literature on political districting shows that the set of criteria most frequently used includes population equality, compactness, and, less often, conformity to administrative boundaries, while the consensus on other criteria is not so widespread. In any case, even when a small number of criteria is taken into account, the main difficulty is due to the fact that they are *conflicting*. A district map with good values for both population equality and conformity to administrative boundaries can perform very badly in terms of compactness; on the other hand, in a very compact district map, the values of conformity and population equality may turn out to be unsatisfactory.

When more than one objective is considered, the solution is not optimal in the traditional meaning of the term. The values that can be obtained for each single criterion in a single-objective procedure are generally better than those obtained for the same criteria in the multiobjective case. Unfortunately, we cannot do any better than finding a good compromise. In the multiobjective case we therefore seek Pareto-optimal solutions. A solution is *Pareto-optimal* if any solution providing a better value for some objective is necessarily worse in terms of at least one other objective. In other words, Pareto-optimal solutions are not uniformly dominated by any other solution.

In general, we find a set of alternative noncomparable solutions, representing different compromises between the criteria. The problem is that we have to choose a unique solution, so some kind of preference order on the criteria must be considered. Often the preference relation can be quantified in terms of weights to assign to the criteria. In such a case, we can define a single objective function that is the weighted combination of the corresponding indicators. Otherwise the choice will be among several Pareto-optimal solutions. This is why a *trade-off analysis* can be very useful when conflicting criteria are considered. The trade-off between two criteria is a quantification of the compromise that we have to accept to avoid situations in which some criterion is largely satisfied and some other criterion is totally unsatisfactory. If we want to satisfy the criteria altogether, we have to find a solution providing good values for all the indicators considered, although we know we cannot reach the optimal values for *all* of them at the same time.

12.2 An Overview of Traditional Models

Traditional districting models are based on a variety of techniques. In the literature there are two main approaches, one based on *division* and the other on *agglomeration*. In the first case, the territory is considered as a whole and the districting procedure works by dividing it into pieces. Since many different cutting strategies can be thought of, many different algorithms can be implemented. In the second case, the territory is viewed as a set T of adjacent territorial units and a district as a subset of such units. In this case the districting technique consists of grouping the territorial units into appropriate subsets of T. Among the cutting strategies, Forrest (1964) suggests operating *successive dichotomies* until the prescribed number of districts is reached. Chance (1965) proposes a *wedge-cutting* strategy that consists of cutting the territory into slices, just as one would do with a cake (in this case each district touches both the center and the boundary of the territory; therefore, little attention is paid to compactness). Another interesting approach is based on *eating up*. Starting from the boundary of the whole territory, one moves toward the center, cutting out a district at each "bite."

The agglomerative techniques include clustering algorithms [Deckro, 1979], multikernel growth techniques [Vickery, 1961; Liitschwager, 1973; Bodin, 1973; Arcese, Battista, Biasi, Lucertini, and Simeone 1992], location [Hess, Weaver, Siegfelatt, Whelan, and Zitlau, 1965; Mills, 1967; Robertson, 1982; Hojati, 1996], and set partitioning techniques [Garfinkel and Nemhauser, 1970; Nygreen, 1988]. Finally, local search can be included in this class of algorithms, because it works on a given partition of the territory by repeatedly rearranging adjacent elementary units. This kind of technique is used by Bourjolly, Laporte, and Rousseau (1981), Browdy (1990), and in the more recent works of Ricca (1996) and Bussamra, França, and Sosa (1996).

Clustering algorithms start by considering the single territorial units as districts and go on by merging them to form new and bigger districts, until the prescribed number of districts is reached. If the number of districts is not fixed in advance, merging stops as soon as a further step turns out to be disadvantageous.

According to the *multikernel growth approach*, a prescribed number of territorial units is selected as the "centers" or the "seeds" of the single districts. The algorithm gradually adds the closest neighboring units to these seeds, until a certain population level is reached. In particular, Vickery (1961) builds one district at a time and, at each stage, he considers a population quota, which is defined as the ratio between the unassigned population and the number of remaining districts.

The use of *location techniques* is a widespread practice in political districting. In fact, this problem can be viewed as a location problem where each unit must be assigned to exactly one center (theoretically, the center of the district), and all the units assigned to the same center form a district. Usually this is done by an iterative location-allocation technique where distance and population criteria are used to aggregate each unit around one of the centers.

The *set partitioning* approach resembles a puzzle that works in two separate and successive phases: first, a large set of potential districts is generated; second, some of them are put together to form a district map where no two districts overlap. The last possibility is to adopt *local search* algorithm, which works by modifying a given initial map locally. This means that only the district boundaries are altered to produce new and better solutions. One can start from a preexisting map or generate a random initial solution and use local search on its own; however, there are authors [Bourjolly, Laporte, and Rousseau, 1981; Arcese, Battista, Biasi, Lucertini, and Simeone, 1992] who prefer to make use of it only as a stage (actually, the final one) of the overall algorithm. See Figure 12.1.

The main districting algorithms are classified in Table 12.1 according to the algorithmic approach they adopt and in Table 12.2 according to the criteria they consider.

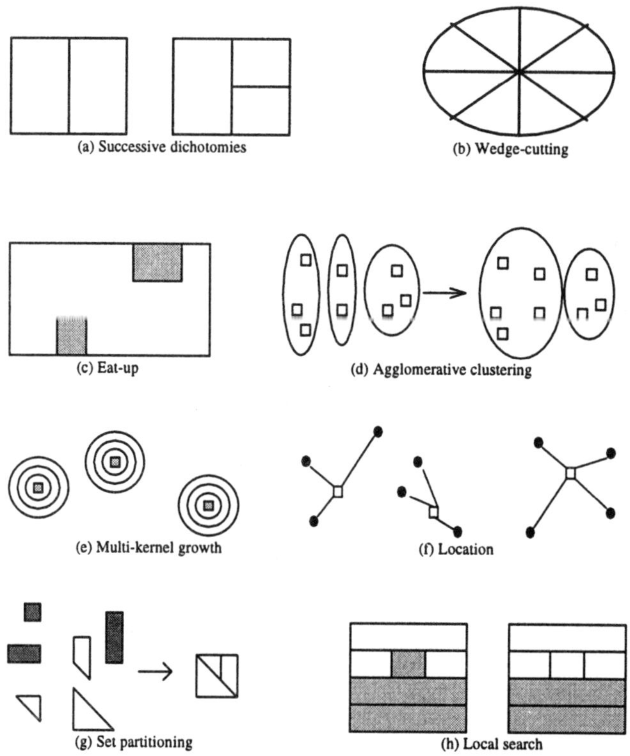

Fig. 12.1. *Algorithmic approaches to districting.*

12.3 Single-Objective Models for Political Districting

The algorithmic approaches mentioned above have been successfully adopted in real applications by several authors. Some well-known models involving only one objective are described in the following paragraphs. Generally, population equality is the objective function. However, this does not mean that single-objective models do not take criteria such as compactness and/or conformity to administrative boundaries into account. In fact, several models involve criteria other than contiguity, absence of holes, and integrity in the form of constraints.

Hess, Weaver, Siegfelatt, Whelan, and Zitlau, 1965

The districting problem is formalized as a discrete location problem. Each territorial unit must be assigned to exactly one center, and all units assigned to the same center form a district. Let n be the total number of territorial units. The objective is to determine the n^2 values of x_{il} that solve the following problem:

Table 12.1. *Districting algorithms classified according to their algorithmic approach (the italicized papers follow a multicriteria approach).*

Algorithmic strategy	***Period*** 60s	70s	80s	90s
Successive dichotomies	Forrest (1964)			
Wedge-cutting	Chance (1965)			
Agglomerative clustering		*Deckro* (1979)		
Multikernel growth	Vickery (1961)	Liitschwager (1973); Bodin (1973)		*Arcese, Battista, Biasi, Lucertini, Simeone* (1992)
Location	Hess, Weaver, Siegfelatt, Whelan, Zitlau (1965); Mills (1967)		Robertson (1982)	Hojati (1996)
Set partitioning		Garfinkel and Nemhauser (1970)	Nygreen (1988)	
Local search			*Bourjolly, Laporte, Rousseau* (1981)	Browdy (1990); *Ricca* (1996); Bussamra, França, Sosa (1996)

$$\min \sum_{i=1}^{n} \sum_{l=1}^{n} d_{il}^2 p_i x_{il},$$

$$\sum_{l=1}^{n} x_{il} = 1, \qquad i = 1, 2, \ldots, n,$$

$$\sum_{l=1}^{n} x_{ll} = k,$$

$$\sum_{i=1}^{n} p_i x_{il} \geq \frac{a}{100} \bar{p} x_{ll}, \quad l = 1, 2, \ldots, n,$$

$$\sum_{i=1}^{n} p_i x_{il} \leq \frac{b}{100} \bar{p} x_{ll}, \quad l = 1, 2, \ldots, n,$$

where x_{il} is a binary variable equal to 1 when unit i is assigned to center l and a and b are the minimum and the maximum allowable district population, calculated as a percentage of the average district population $\bar{p}$. Moreover, notice that the variable x_{ll} takes the value 1 whenever unit l is chosen as one of the centers.

The first n constraints mean that each unit must belong exactly to one district; the next constraint means that the total number of districts must be exactly k. The other two groups of n constraints represent the conditions on the maximum and the minimum allowable district population, expressed as a

Table 12.2. *Districting algorithms and criteria.*

Algorithm	***Criteria***				
	Contiguity	Absence of holes†	Population equality	Compactness	Conformity
Vickery	*	*	*	*	*
Forrest	*	*	*	*	
Chance	*	*	*		
Hess et al.	*		*	*	
Mills	*		*	*	
Deckro	*		*	*	
Garfinkel, Nemhauser	*		*	*	
Liitschwager	*	*	*	*	*
Bodin	*		*		
Bourjolly et al.‡	*		*	*	
Robertson	*		*	*	
Nygreen	*		*	*	*
Browdy	*		*	*	
Arcese et al.	*	*	*	*	*
Ricca	*	*	*	*	*
Hojati	*		*	*	
Bussamra, França, Sosa	*	*	*	*	*

†In this column we mark only the cases in which this criterion is explicitly stated, but it is clear that whenever compactness and contiguity are combined together, and the outcome is a sufficiently compact plan, then the absence of holes (or enclaves) is also satisfied.

‡Remember that this algoritm considers a socioeconomic homogeneity criterion.

percentage of $\bar{p}$. Finally, the objective function is a measure of compactness (total inertia), while population equality is taken into account in the inequality constraints.

The algorithm is an iterative heuristic and, at least in theory, its convergence is not guaranteed. Essentially the algorithm consists of six steps:

(1) guess the district centers;

(2) use a transportation algorithm to assign population equally to these centers at minimum cost (defined as the minimum sum of squared distances d_{il}^2 of each unit from the center of its district);

(3) adjust assignment so that each territorial unit is entirely within one district;

(4) compute centroids and use them as improved district centers;

(5) repeat from step 1 until solution converges;

(6) try more initial guesses.

In step 2 a transportation algorithm is used to assign population equally to the centers. This often causes a territorial unit to be assigned to more than one district so that a splitting problem must be solved in order to eliminate multiple assignments. In practical applications, it has been observed [Hess, Weaver, Siegfelatt, Whelan, and Zitlau, 1965] that the heuristic converges to a local minimum in less than ten iterations (ten transportation problems must be solved).

Garfinkel and Nemhauser, 1970

This is a two-phase algorithm, which takes into account three basic criteria: population equality, compactness, and contiguity. All these criteria are constraints during Phase I. In Phase II only population equality is considered, and it is taken as the objective function.

All possible feasible districts are generated in Phase I. Such districts have a population that differs from the average district population $\bar{p}$ by at most α times $\bar{p}$, where α is a given parameter between 0 and 1, i.e., $|P_j - \bar{p}| \leq \alpha\,\bar{p}$. Moreover, they must satisfy two compactness constraints. First, the distance between any pair (i,l) of units belonging to the same district cannot exceed a prescribed *exclusion* distance e_{il}. In addition, for each feasible district j, compute the farthest distance d_j between two units belonging to district j and the area A_j of district j. Then the ratio $\frac{d_j^2}{A_j}$ cannot be greater than a given threshold β. This is a measure of the compactness of district j based on its shape.

In Phase II a district map is built by the selection of a set of k disjoint feasible districts that covers all territorial units and such that the maximum population imbalance is minimized. The model adopted is the following:

$$
\begin{aligned}
\min \max & \left\{ \tfrac{|P_j - \bar{p}|}{\alpha \bar{p}} x_j : j = 1, 2, \ldots, K \right\}, \\
& \sum_{j=1}^{K} a_{ij} x_j = 1, \qquad i = 1, 2, \ldots, n, \\
& \sum_{j=1}^{K} x_j = k, \\
& x_j = 0, 1, \qquad j = 1, 2, \ldots, K,
\end{aligned}
$$

where K is the number of feasible districts generated during Phase I and $a_{ij} = 1$ if unit i belongs to feasible district j, and 0 otherwise. Here x_j is equal to 1 when district j is selected, and to 0 otherwise.

Nygreen, 1988

Consider a connected graph G, where the nodes represent territorial units and an arc between two nodes exists if and only if the corresponding units have common boundary parts. The districting problem can be stated as follows: delete edges from the graph until it becomes a forest with as many trees as the given number of districts. In such a case, contiguity is a consequence of the connectedness of the trees, while compactness can be obtained by requiring that each tree must have maximum depth two (this means that it is always possible to choose a node c—called the *root* of the tree—such that any other node is adjacent either to c or to some node adjacent to c). Nygreen also adopts some kind of conformity criterion (which he calls "political constraint") according to which all territorial units in the same city must belong to the same district. Population equality is the objective of this model, and it is formalized as the sum of squared deviations of district populations from the mean $\bar{p}$. This function is replaced by a piecewise linear approximation with a small number of breakpoints, which obviously has to be minimized. Nygreen's algorithmic approach is very similar to the one proposed by Garfinkel and Nemhauser (1970). The algorithm consists of two stages: in the first one, all feasible districts are generated; in the second one, districts are combined to form a district plan. The main contribution of Nygreen's work is to be found in the first stage: two important rules, the *root rule* and the *residual graph test*, are used to reduce the number of roots at the beginning of the algorithm and to generate the feasible districts, respectively. For each tree d of maximum depth two (which represents a potential district d) the residual graph test makes it possible to verify whether d is a valid district or not. Given a district d, the *residual graph* (or, in Nygreen's terminology, the "graph that remains") is the subgraph obtained from G after deletion of the nodes in d and their incident edges. Given that the total number of districts is equal to k, district d is *valid* if it is possible to find in the residual graph other $k-1$ districts that form, together with d, a feasible district plan. Once the set of all possible districts is defined, a set partitioning problem can be formulated. While Garfinkel and Nemhauser use an implicit enumeration technique to solve this problem, Nygreen uses standard integer programming software.

Hojati, 1996

Like Hess, Weaver, Siegfelatt, Whelan, and Zitlau (1965) Hojati adopts a location approach, but some significant differences can be observed. In the first phase of the algorithm, the centers of the districts are located and the territorial units are assigned to the centers by a transportation technique (a territorial unit can be assigned to more than one center). Then a sequence of capacitated transportation problems is solved to force the assignment of each territorial unit to only one center. The main difference with respect to Hess, Weaver, Siegfelatt, Whelan, and Zitlau (1965) is in the choice of the district centers. Instead of adopting an iterative strategy based on successive adjustments, Hojati locates the centers at the beginning of the procedure, and this choice is permanent. The problem is formulated as the following mixed integer program:

$$
\begin{aligned}
\min \quad & \sum_{i=1}^{n}\sum_{l=1}^{n} c_{il} X_{il}, & \\
& \sum_{l=1}^{n} X_{il} = 1, & i = 1, 2, \ldots, n, \\
& \sum_{i=1}^{n} p_i X_{il} = \bar{p} Y_l, & l = 1, 2, \ldots, n, \\
& \sum_{l=1}^{n} Y_l = k, & \\
& 0 \leq X_{il} \leq Y_l, & i, l = 1, 2, \ldots, n, \\
& Y_l = 0, 1, & l = 1, 2, \ldots, n,
\end{aligned}
$$

where n is the number of territorial units, k is the number of districts, p_i is the population of unit i, and c_{il} is the squared Euclidean distance between unit i and unit l, multiplied by p_i. The variable X_{il} is the fraction of the demand (as given by the population) of unit i charged to center l, while Y_l is a binary variable equal to 1 when unit l is chosen as the center of a district, and 0 otherwise.

A Lagrangian relaxation of the problem is derived and then transformed into a zero/one program which is solved by a subgradient optimization algorithm (for a review of these methodologies see, e.g., Nemhauser and Wolsey, 1988). Once a solution $(Y_1, Y_2, \ldots, Y_n)$ has been found, it defines the set of territorial units that have been selected as the centers of the k districts. Now it is necessary to aggregate the other units. In order to deal with this phase, a transportation problem is solved: each territorial unit is an origin and each center a destination; the supplies are equal to the populations of the units and all demands are equal to $\bar{p}$; the transportation costs are defined on the basis of both the distance between each unit i and each center l and the population of unit i. Thus, there are $k\,n$ variables X_{il} and $k+n$ equations, and it can be shown that the number of units split between two or more districts in the solution of this transportation problem is at most $k-1$.

Now a *split-resolution problem* (SRP) must be solved. Let I' be the sets of split units; L' is the set of those centers l such that some split unit $i \in I'$ has been (partially) assigned to l. Let X^*_{il}, $i = 1, 2, \ldots, n$; $l = 1, 2, \ldots, k$, be the solution of the transportation problem solved in the previous phase. We consider $I' \cup L'$ as the set of nodes of a graph F where an arc (i, l) exists if and only if $X^*_{il} > 0$. The SRP can be formulated as follows:

$$
\begin{aligned}
& \text{minimize the number of arcs } (i, l) \text{ in } F \text{ with } x_{il} > 0, & \\
& \sum_{l \in U_i} x_{il} = p_i, & i \in I' \\
& A_{\min} \leq a_l + \sum_{i \in U_l} x_{il} \leq A_{\max}, & l \in L', \\
& x_{il} \geq 0, & i \in I', l \in L',
\end{aligned}
$$

where $A_{\min}$ and $A_{\max}$ represent the minimum and the maximum allowed population in each district, while U_v is the set of nodes adjacent to node v in F, a_l is the size (population) of district l without split units, and here the variable x_{il} represents the amount (rather than the fraction as above) of population of unit i charged to center l. Hojati proves that SRP is computationally hard to solve and he suggests a heuristic.

12.4 Multiobjective Models for Political Districting

Multiobjective optimization models take into account more than one objective at a time. There are different ways to cope with multiobjective models: a common strategy is to consider the different criteria according to a fixed hierarchy reflecting a specific preference order; another strategy is to build a mixed objective function combining all the objectives into one. In the following sections we will describe the main multicriteria applications to the districting problem, focusing both on the models adopted and the algorithms designed. Of course, the problem is difficult, so using heuristics is the best we can do.

Deckro, 1979

Deckro suggests a multiple objective heuristic, which is based on a clustering technique. The different objective functions $p_h(d)$ represent the "value" of district d (regarded as a set of elementary territorial units) according to criterion h. They are all assumed to increase when a unit is added to d, and are ordered with respect to the importance of the relative criteria. A target value is prescribed for each objective together with a (not necessarily symmetric) range of acceptable variation, and the aggregation of the elementary territorial units takes into account these target values.

The criteria are considered one at a time, starting from the most important, and the districts are sorted in ascending order according to it. At each iteration a set of districts is under construction. These districts are ranked according to the most important objective (say, objective 1). The district with the smallest value, say, district d, is selected and the algorithm searches for an adjacent district that can be added to d to improve objective 1. When the list of adjacent districts (the *candidate list*) is scanned, three different situations can arise:

(1) adding any candidate of the list to d results in an upper bound violation for objective 1;

(2) adding any candidate of the list to d results in a lower bound violation for objective 1;

(3) there is at least one candidate who can be added to d such that the value of objective 1 is within the acceptable range of variation.

In case 1 district d is removed from the set of districts under construction and considered as a completed district. Case 2 means that even if a district is added to d, the new district is still under construction; therefore, the candidate with the smallest value for objective 1 should be added to d. The rank must be updated considering the new value for this district. If case 3 occurs and there is just one candidate who can be added to d, then the aggregation is performed and the rank updated. Otherwise, if there is more than one candidate who can be added to d without violating the thresholds for objective 1, the second most important objective is considered. Again, one of the three cases mentioned above must hold. If case 1 holds the candidate with the smallest value for objective 2 is added to d; in the other two cases the aggregation is performed as described above, repeating the whole procedure with the third, fourth, etc., objective when necessary. The procedure stops when the candidate list is empty. Following this procedure, the final solution may violate both upper and lower bounds, especially for the less important objectives. Actually, it can occur that some objective is never considered in the procedure because case 3 may not arise so often. Nevertheless, the advantage of such a procedure is that the effects of different preferences can be tested by altering the range width of the targets and the ordering of the objectives, so that the final selection is left to the decision maker.

Bourjolly, Laporte, and Rousseau, 1981

The Bourjolly, Laporte, and Rousseau algorithm has been designed to develop a districting plan for the Ile de Montréal in 1979. Population equality, compactness, and socioeconomic homogeneity are considered as objectives. Although socioeconomic homogeneity is usually criticized because it does not satisfy the neutrality condition, in this case it may be considered essential to guarantee the representation of the multiethnic composition of the population under study.

A traditional districting problem is considered with binary variables and constraints on contiguity, the number of districts, and the maximum allowed deviation from average district population. Criteria such as conformity to natural barriers and administrative boundaries are also considered. There is a single objective function f obtained as weighted average of the three objectives mentioned above. The algorithm is a heuristic that at each iteration reduces the value of f. The special feature of this algorithm is the preprocessing phase adopted to build a list of candidate districts for each territorial unit. More precisely, at the beginning the list of candidate districts for unit i contains only the district whose center is nearest to i. At each iteration, after having eliminated all districts whose centers fall within a prescribed angle ϑ, bisected by the line through i and the center of the last chosen district, one adds to the list the noneliminated district whose center is nearest to i (see Figure 12.2).

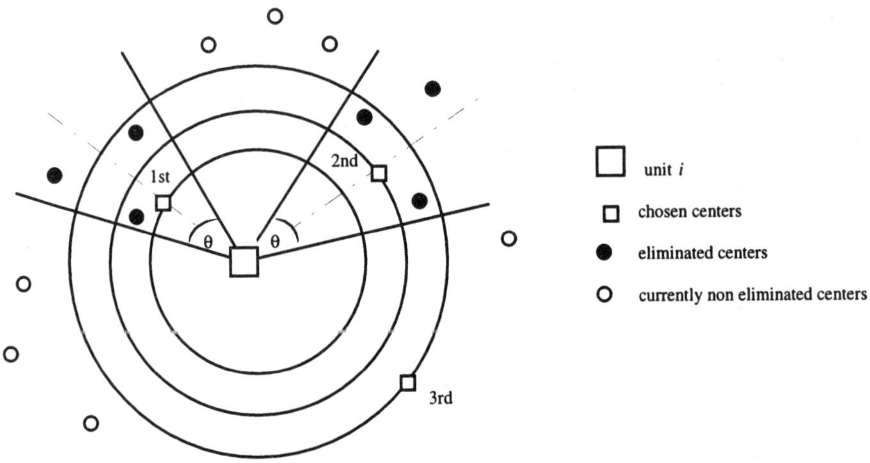

Fig. 12.2. *Selection of centers to be included in the list of unit i.*

Arcese, Battista, Biasi, Lucertini, and Simeone, 1992

Another approach is based on building graph theoretic models for political districting. In fact, in many territorial problems a graph representation model is useful. Given a territory (for example, a region, a city, etc.) partitioned in contiguous elementary territorial units (for example, counties, wards, etc.), a connected graph $G(N, E)$ can be defined as follows: the set of nodes N corresponds to the set of elementary units, and there is an edge between two nodes $i, j \in N$, if they correspond to adjacent territorial units (if two units touch just in one point, they are not considered to be adjacent).

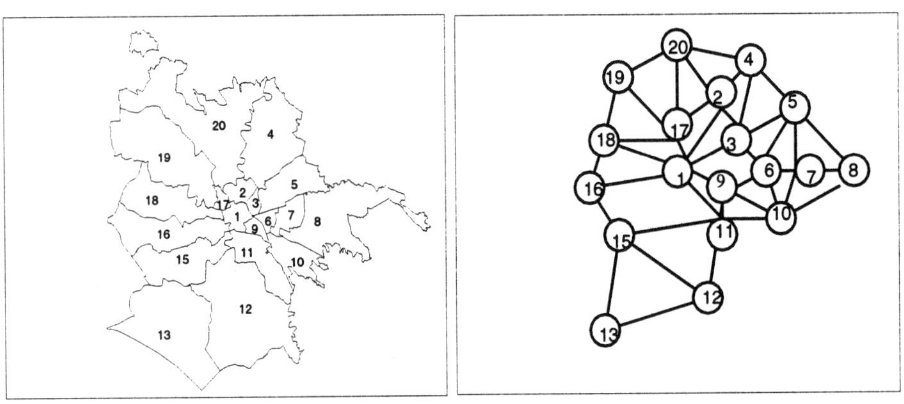

Fig. 12.3. *Map and graph of the city of Rome divided into wards.*

Once the graph $G(N, E)$ has been constructed, a district can be viewed as a subset of nodes in N that induces a connected subgraph of G. For example in Figure 12.3 the subset of nodes {1,3,6,7,9} induces a connected subgraph, but this is no longer true if node 20 is added to the subset. A district map can be viewed as a partition of the node set into connected components. Furthermore, the population (or the number of electors) of a territorial unit can be associated with the corresponding node as a weight, and the distances between pairs of units can be adopted as weights for the edges in E (generally, in real applications road distances are considered). These weights are necessary to calculate the value of the criteria indicators. According to the definitions given above, the political districting problem can be stated as follows: find a partition of the graph representing the territory into a prescribed number of connected components so as to maximize (in the vector maximization sense) a given set of objectives.

On the basis of this kind of representation, Arcese, Battista, Biasi, Lucertini, and Simeone (1992) suggest an algorithm with a multiple objective function where population equality, compactness, and conformity to administrative boundaries are taken into account simultaneously.

To be more specific, consider a connected graph $G(N, E)$. Let $n = |N|, m = |E|$, and let k be an integer such that $1 \leq k \leq n$. A partition $\pi = \{C_1, C_2, \ldots, C_k\}$ of N into k subsets (briefly, a k-partition) is said to be *connected* if, for each $j = 1, 2, \ldots, k$, the subgraph $G(C_j)$ induced by C_j is connected. The set of all connected k-partitions will be denoted by $\Pi_k(G)$. Then the districting problem can be formulated as the vector maximization problem

$$\max_{\pi \in \Pi_k(G)} \{f_1(\pi), f_2(\pi), f_3(\pi)\},$$

where $f_1(\pi), f_2(\pi), f_3(\pi)$ measure population equality, compactness, and conformity to administrative boundaries, respectively. The above vector maximization problem consists of finding a partition $\pi^* \epsilon \Pi_k(G)$ that is Pareto optimal with respect to the three objectives. Arcese, Battista, Biasi, Lucertini, and Simeone (1992), who apply their algorithm to the Italian case, further assume that the territory is divided into regions (each unit belongs to one region) and prescribe that no district should be split between two or more regions: this requirement is dictated, of course, by conformity to administrative boundaries. The algorithm consists of two stages: a regional optimization and a national optimization stage. First, for each region R, compute the number of districts to be assigned to that region. Let q_R be the quotient $\frac{P_R S}{P}$,where P_R, S, and P are the population of the region, the total number of seats, and the total population, respectively. When q_R is an integer, we can assign exactly q_R seats to region R. Otherwise (q_R is not an integer) the number of districts is chosen as one of the two integers $\lfloor q_R \rfloor$ and $\lceil q_R \rceil$. In this way, one achieves an equitable assignment of the seats to the regions. "Equitable" means that the *Hare property* mentioned in section 5.2 is satisfied.

In the regional optimization stage two district plans are calculated for each region: one with exactly $\lfloor q_R \rfloor$ districts and the other with $\lceil q_R \rceil$ districts (the two plans coincide if q_R is an integer). Both plans are required to be "good" with respect to all the criteria considered.

Once the two plans have been obtained for each region, a global optimization stage is carried out by considering simultaneously all the regions and choosing for each region R one of the two district maps, such that the total number of districts is equal to k and a global performance index is maximized (national optimization). Technically, this second phase requires the solution of a cardinality-constrained linear binary knapsack problem via a greedy algorithm as explained in section 6.3.

Let us now outline the main stages of the algorithm. Recall that each region is represented by a connected graph and each node of the graph is an elementary unit, while each edge links two adjacent units.

(1) *Preliminary choice of seeds.* The first step of the procedure is the choice, in the region under consideration, of a subset of nodes, which are candidates to become the seeds in the multikernel growth procedure of stage 3. Typically, if the number of districts in the region R is k_R, the subset contains more than k_R candidate seeds (say, about $2k_R$), so as to allow for some flexibility in the final choice of the seeds, one per district, to be carried in stage 2. The candidate seeds are selected far from each other and well distributed over the territory. Usually, they are chosen among units with largest populations. If there are not enough high population units to choose from, low population candidate seeds can be selected under the condition that they are not adjacent to any other low population seed. These requirements make it easier to build feasible districts.

(2) *Location-allocation algorithm.* For any given choice of k_R seeds from the candidate set, the *influence zone* related to each seed in R is defined as the set of nodes that are closer to such a seed than any other. This stage consists of an iterative *location-allocation* procedure[9] which, starting from an initial choice of k_R seeds, gradually modifies the choice of the seeds until each seed is placed in the center of its influence zone. A further adjustment may be necessary to balance population.

(3) *Multikernel growth algorithm.* All districts initially consist of a single seed. Each district D carries an adjacency list referring to the nodes that may by included in D. Therefore, these nodes are not contained in D but are adjacent to it.

Then the districts are considered one at a time, and they are completed only when the corresponding adjacency list is empty. At each step, the district with the lowest population is selected and a unit in the adjacency list is added such that a suitable function of the three criteria is optimized.

(4) *Descent algorithm.* The map obtained at the end of phase 3 can be further improved by working on the units that lie on district boundaries. In this phase, the algorithm seeks better solutions by moving nodes from a district to a neighboring district and considering a linear combination of the single indexes with fixed weights.

There are some other features that take into account further requirements. In particular, the algorithm treats cities with more than 140,000 inhabitants

[9]Such procedure is well known in location theory (see, e.g., Kuehn and Hamburger, 1963; Cooper, 1963; Maranzana, 1964), as well as in cluster analysis (Forgy, 1965; MacQueen, 1967) where it is known under the name of *k-means*.

1. CITY IDENTIFICATION
   Townships with at least 140,000 inhabitants are cut out from the regions they belong to, and they are treated like separate regions.

2. REGIONAL OPTIMIZATION
   For each region R:
   - 2.1 Calculate the number of districts in region R; if $q_R = P_R \frac{S}{P}$ the number of districts is either $\lfloor q_R \rfloor$ or $\lceil q_R \rceil$; for $k_R = \lfloor q_R \rfloor, \lceil q_R \rceil$ determine a district map for region R in k_R districts performing the following steps:
     - 2.2 preliminary choice of district seeds;
     - 2.3 location-allocation algorithm for the final choice of seeds;
     - 2.4 multikernel growth algorithm to find an initial distriction map;
     - 2.5 descent algorithm to find a final district map.

3. NATIONAL OPTIMIZATION
   Solve a cardinality-constrained linear binary knapsack problem using a greedy algorithm to define the number of the districts ($\lfloor q_R \rfloor$ or $\lceil q_R \rceil$) for each region R such that a global performance indicator is maximized.

Fig. 12.4. *Main steps of the algorithm in Arcese, Battista, Biasi, Lucertini, and Simeone* (1992).

separately, partitioning them into smaller units just as if they were regions. In these cases, smaller elementary units (wards) are considered, and in the regional optimization phase each such city is cut out from its region and considered as an additional region.

Figure 12.4 shows the main steps of the algorithm. Notice that this algorithm concentrates on preliminary phases providing a very good starting point, sufficiently near to the final solution. In fact, the amount of time spent to choose the initial seeds is much more than that needed by other algorithms that use random choice, but this allows us to build "good" districts from the beginning.

12.5 Local Search Algorithms for Political Districting

On the basis of the graph-theoretic model suggested in Arcese, Battista, Biasi, Lucertini, and Simeone (1992), several local search heuristics for the districting problem (namely, the simple descent algorithm, simulated annealing, tabu search, and old bachelor acceptance) have been tested on a sample of five Italian regions, providing comprehensive comparative numerical results (which are generally lacking in the literature on districting algorithms) [Ricca, 1996; Ricca and Simeone, 1997].

Population equality, compactness, and conformity to administrative boundaries are the objective functions, while integrity, contiguity, and absence of holes are the constraints that are automatically satisfied by the graph model and the way the heuristics work. For a given territory a series of runs are performed, each one with a different objective function. In particular, three "pure" objective functions are considered with respect to each of the three criteria mentioned above, and a fourth objective function is defined as a weighted linear combination of the other three, with weights in the [0,1] interval. Local search methods are useful to find near-optimal solutions to hard combinatorial problems. These procedures try to improve a given solution by testing the variation of the objective function when small perturbations are operated on the current solution. The perturbations are *local* because they involve only moves from a solution s to another s', which lies in a suitably defined "neighborhood" of s.

Consider a given minimization problem. The simplest version of local search (descent algorithm) works by performing a modification of the current solution only if this generates a new solution that is better (lower values for the objective function) than the immediately previous one. There are other, more sophisticated, local search algorithms that allow the objective function to worsen temporarily to avoid being trapped in bad local minima.

Since we are referring to heuristics, the final solution will generally be a "good" local optimum. In fact, Pareto optimality is hard to test and hence not very useful. On the other hand, locally Pareto-optimal solutions can be defined as follows: given a multiobjective optimization problem with r objectives and assuming that for each solution a neighborhood of "nearby" solutions is defined, a solution s is *locally Pareto-optimal* if no local move can improve one of the objectives without worsening at least one of the other $r-1$. Notice that, if the neighborhoods are "small," local Pareto optimality becomes computationally much easier to check than Pareto optimality.

According to the graph model described above, a local move corresponds to a *migration* of a unit i from a district d to a neighboring district d'. Furthermore, a migration is *feasible* if a node adjacent to i exists and moving i out of its district of origin d does not disconnect d itself. More formally, consider the graph $G(N, E)$ and a partition $\pi \in \Pi_k(G)$; the migration moves a node from a class C_p of π to a different class C_q of π. Obviously, this operation transforms π into a new partition $\pi' \in \Pi_k(G)$, which differs from the original one just for classes C_p and C_q.

The example in Figure 12.5 shows feasible and infeasible migrations on a simple connected graph. Starting from the connected 3-partition (a), the move involves districts d_1 and d_3 and node 6, which migrates from d_1 to d_3 (and this is possible because 6 is adjacent to 5). This is a feasible migration because the resulting partition (b) is still connected. On the other hand, starting again from (a), the migration of node 3 from d_3 to d_1 is infeasible, since the resulting partition (c) is no longer connected. Notice that node 3 cannot be moved to any other district because it is crucial for the connectivity of district d_3.

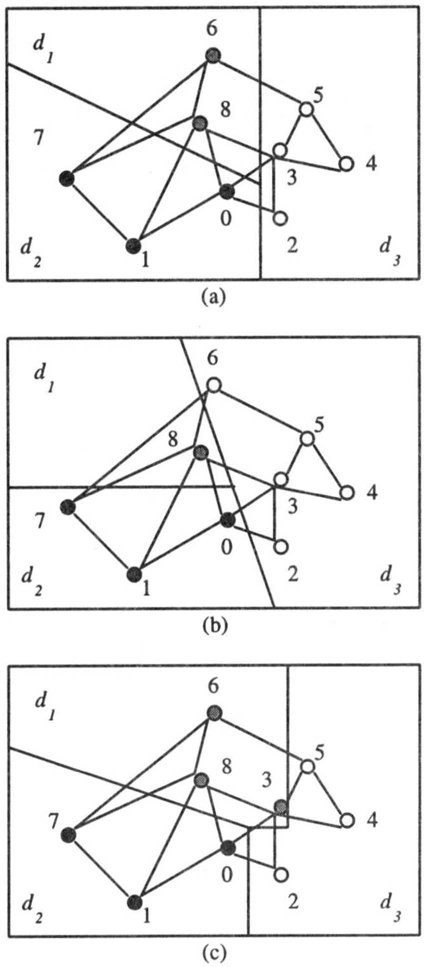

Fig. 12.5. (a) *The initial partition;* (b) *a feasible migration;* (c) *an infeasible migration.*

Notice that, starting from a partition without holes and noncontiguous districts, a feasible move always preserves these two properties; therefore, these local search procedures automatically generate district plans that always satisfy both these constraints, provided that the initial plan does so.

Now a general scheme of the descent algorithm can be sketched out using the notion of feasible migration (Figure 12.6).

The descent algorithm changes the current solution only when the objective function improves, and it stops when the set of feasible migrations has been completely explored. This is an important drawback that prevents the algorithm from finding more than one local optimum. In fact, when all the boundaries of districts in the map have been checked, all the directions for the search

```
Descent algorithm

1.    Select an initial solution s;
2.        While some feasible migration from s exists, do:
2.1.          try to perform one of the feasible migrations, obtaining
              a new solution s'. Let Δ = f(s') − f(s) be the variation
                    of the objective function when going from s to s';
2.2.          if Δ ≤ 0:
                    the migration is performed;
2.3.          else [Δ > 0]:
                    the migration is rejected;
              endif;
3.        endwhile
4.    Output the best solution s found up to now.
```

Fig. 12.6. *Simple descent algorithm.*

are precluded and the algorithm cannot proceed any further. However, the temporary worsening of the objective function can be a good way to avoid the algorithm becoming trapped in a bad local minimum. What an efficient local search heuristic should do is to guide the search toward local minima, recognizing the bad ones and changing the direction of the search when it reaches a point that is too close to a bad local minimum. In this case, when the algorithm stops, a good local optimum has been reached. Generally, the value of the solution found will not coincide with the global optimum, but it may happen that they coincide since the global optimum is a local optimum itself. The rationale underlying these heuristics is to follow a longer search path with respect to the simple descent algorithm in order to find better values for the objective. For example, consider a problem where the objective function $f(x)$ shown in Figure 12.7 must be minimized. Both points B and A are local optima, but B is better than A because the value of the objective function is lower. Assume that point C represents the initial solution. If the sequence of migrations that will be performed leads the search toward point A, then the algorithm stops in A because any migration from A to a nearby point produces a disadvantageous solution ($\Delta > 0$), and point B will never be touched. On the other hand, if the objective function is allowed to worsen, the search will continue even if it reaches A. In fact, even if point A is reached, one can still proceed to point D by a temporary worsening and explore more than one local optimum. In this case point B will be the final solution.

The main problem with these heuristics is *cycling*. In fact, since both improvements and worsening are allowed, the same sequence of migrations might be infinitely repeated. Each algorithm has its own strategy to prevent cycling. Efficient search procedures are inspired by processes arising in other scientific branches, such as biology, physics, and computer technology. For example, simulated annealing is based on the physical annealing process, and tabu search is

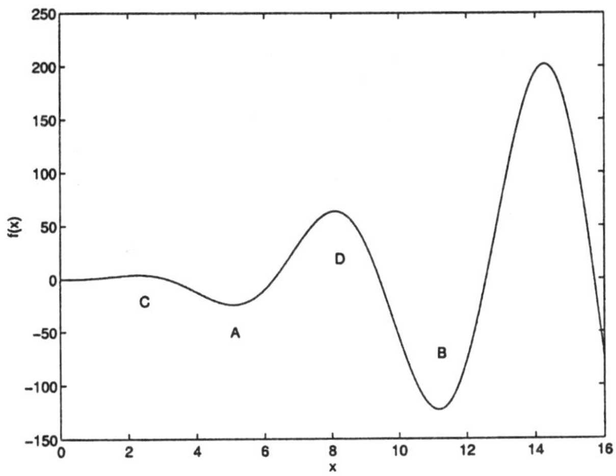

Fig. 12.7. *A and B are both local optima, but B provides a better objective value than A.*

based on the processes of artificial intelligence. Let us briefly analyze simulated annealing, tabu search, and old bachelor acceptance by sketching a scheme for each algorithm and giving some further insights. See Figures 12.8, 12.9, and 12.12.

Simulated annealing is not very different from the descent algorithm, but additional parameters are introduced to define the probability of performing a disadvantageous migration and to determine the stopping condition of the algorithm. The parameters T_0, T_f, and r represent the *initial temperature*, the *final temperature*, and the *cooling rate*, respectively. Once values are fixed for T_0, T_f, and r (T_0 and T_f are positive values while r is in $[0, 1]$), at each iteration the temperature is decreased by a factor r (T is replaced by rT) and a migration is performed with probability 1 if it improves the objective function, while it is performed with a certain probability (generally equal to $e^{-\Delta/T}$) if it produces a worsening of the objective function. The traditional definition of the probability that is adopted here is a decreasing function of the temperature so that, when only a few iterations are left, the algorithm favors advantageous migrations to reach the nearest local optimum. Furthermore, at a fixed temperature, such probability favors migrations that produce a small worsening of the objective function. This procedure tends to avoid bad local minima and cycling: in fact, the slower the cooling, the better the sequence of local optima found. The algorithm stops when the temperature reaches its final value T_f or when the maximum number L of iterations has been reached.

The basic feature of tabu search is the use of *tabu moves*: when a move from s to s' is performed, the reverse move from s' to s is forbidden for a certain number of iterations. Figure 12.10 shows two neighboring districts where the

Simulated annealing [Kirkpatrick, Gelatt, and Vecchi, 1983; Cerny, 1985]

1. Select an initial solution s;

2. Let $T = T_0 > 0$ be the initial temperature.

3. While $T \geq T_f$, do:

3.1. for $i = 1, 2, \ldots, L$, do:

3.1.1. try to perform a feasible migration from s, obtaining a new solution s'. Let $\Delta = f(s') - f(s)$ be the variation of the objective function when going from s to s';

3.1.2. if $\Delta \leq 0$ (downhill move):
the move is performed;

3.1.3. else $[\Delta > 0]$(uphill move):
the move is performed with probability $e = e^{-\frac{\Delta}{T}}$;
endif

3.2. update $T = rT$;

3.3. endfor;

4. endwhile.

5. Output the best solution s found up to now.

Fig. 12.8. *Simulated annealing.*

Tabu search [Glover, 1989 and 1990]

1. Select an initial solution s;

2. Do:

2.1. generate a set of feasible migrations from s;

2.1.1. if there are nontabu moves that produce a new solution s' such that $\Delta = f(s') - f(s) \leq 0$:
then perform the one with the largest absolute value of Δ;

2.1.2. else [every nontabu move produces a new solution s' such that $\Delta = f(s') - f(s) > 0$] :
perform the one that produces the lowest $\Delta > 0$;
endif;

2.2. update the tabu list;

3. until the stopping condition is satisfied.

4. Output the best solution s found up to now.

Fig. 12.9. *Tabu search.*

black nodes are the units in d' and the white nodes are the units in d. Node v is on the boundary of district d, and by migrating from d to d' it cannot disconnect it. After the migration (Figure 12.11), node v is in district d', but nothing prevents the search technique from selecting it again to perform the reverse migration (in the next iteration or some later iteration) since it still is feasible. However, this migration should not be performed because it brings the partition back to the original one shown in Figure 12.10. Obviously, this should be avoided to prevent the algorithm from cycling and from exploring several times a restricted part of the feasible region.

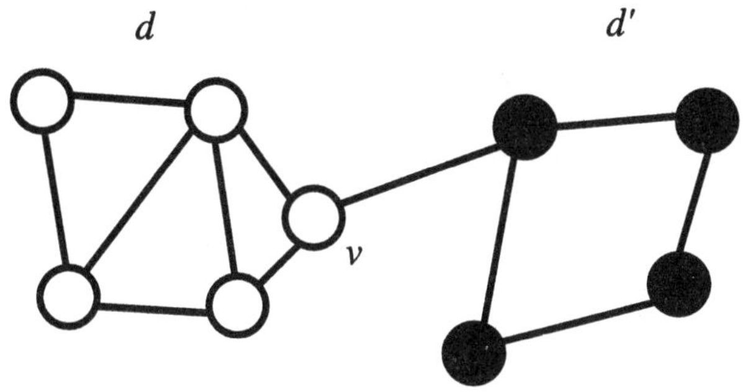

Fig. 12.10. *v is a neighboring node between district d and district d'.*

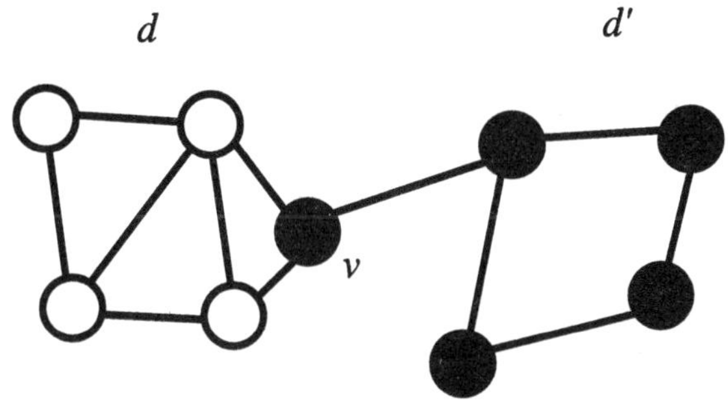

Fig. 12.11. *Districts d and d' after the migration of v.*

The tabu list is the set of tabu moves. It is generally encoded as a circular list (with fixed or variable size) so that a move that is recorded as a tabu move can go back to the set of feasible moves after a (fixed) number of iterations. The algorithm selects nontabu moves that improve the solution, but if there are none left, it prefers to perform a nontabu disadvantageous move rather than a tabu one, even if this locally worsens the objective function.

Tabu search does not include a proper stopping condition. Traditional rules can be adopted, such as fixing the total number of iterations, or the maximum number of successive worsening steps allowed, etc., or by combining these rules together.

Old bachelor acceptance is a threshold acceptance method with a special threshold adjusting mechanism. In fact, at each iteration there is a threshold value defining the maximum acceptable worsening for the objective function.

This implies that the objective may improve, but it also may worsen, up to a certain limit. At each iteration, the threshold adjusts itself following a *non-monotone schedule*, and even negative threshold values can be reached. In particular, the threshold decreases each time the algorithm improves the current solution, while it increases when a disadvantageous step is performed. This strategy allows us to avoid bad premature local optima, and it enables us to find new descent directions when we are very far from the optimum. The strategy of this algorithm consists of alternating *dwindling expectation* (to escape a local minimum) to *ambition* (to explore the solution space and find new local minima). The main steps of the old bachelor procedure are shown in Figure 12.12, where $\Delta^+(i), \Delta^-(i) > 0$ are the two functions used to update the threshold and m is the total number of iterations. Some preliminary experiments indicate a marked superiority of old bachelor acceptance over the other heuristics.

Old bachelor acceptance [Hu, Kahng, and Tsao, 1995]

1. Select an initial feasible solution s;
2. Fix an initial threshold T_0.
3. For $i = 1, \ldots, L$:
3.1. try to perform a feasible migration from s obtaining a new feasible solution s'. Let $\Delta = f(s') - f(s)$ be the variation of the objective function when going from s to s';
3.2. if $\Delta \leq 0$ (downhill move):
    the move is performed;
    $T_{i+1} := T_i - \Delta^-(i)$;
3.3. else $[\Delta > 0]$ (uphill move):
    the move is not performed;
    $T_{i+1} := T_i + \Delta^+(i)$;
  endif;
 endfor.
4. Output the best solution s found up to now.

Fig. 12.12. *Old bachelor acceptance.*

12.6 Trade-offs between Criteria

The aim of a districting algorithm is to find, within a reasonably short time, solutions that realize a good compromise among the criteria. At the moment no efficient algorithm for finding Pareto-optimal solutions for the districting problem exists. The available ones can just find a set of "good" solutions, that is, solutions with good values for all criteria (population equality, compactness, conformity to administrative boundaries, and so on). Pareto optimality is something very difficult to achieve because these criteria are structurally in conflict, and even optimizing only one of them is computationally hard [Altman, 1997].

Many examples of conflict between reasonable criteria are scattered in real political districting applications. One can easily imagine the conflict arising between integrity, which usually makes districts very irregular, and compactness, which requires circular zones. This makes the evaluation of trade-offs a fundamental aspect when a multicriteria approach is adopted. Typically (e.g., in interactive methods for multicriteria optimization), trade-offs are analyzed when the decision maker is faced with the choice between two different alternatives. Trade-offs are measured in terms of (absolute or relative) variations of the criteria values from one solution to the other. In order to analyze the compromise (or trade-off) between criteria we suggest a graph representation that allows us to relate each objective function to the others. We are especially interested in the comparison between each objective function and a *target* objective. For r objectives, a *trade-off graph* is defined as an r-node, directed, and complete graph. Each node represents one objective and the $r(r-1)$ arcs connect pairs of criteria. Both the nodes and the arcs of the trade-off graph carry weights. Given two arbitrary district plans π and π', the weight of node h is defined as $w_h = f_h(\pi') - f_h(\pi)$, where f_h is the hth objective; the weight of arc (h, l) is $u_{hl} = \frac{w_l}{w_h}$ (if $w_h = 0$ then u_{hl} is undefined). Notice that both node- and arc-weights may take positive, null, or negative values.

Given an initial (random) district map, it is easy to visualize on a trade-off graph whether there has been a gain or a loss in the corresponding objective after the optimization phase. We only have to read the trade-offs between the criteria on the arcs.

To better understand the possible use of a trade-off graph, we report some examples. Figures 12.13, 12.14, and 12.15 show the trade-off graphs corresponding to a recent application of local search techniques to the districting problem in Italy [Ricca, 1996]. Traditional criteria (that is, population equality, compactness, and administrative conformity) and the tabu search technique have been considered in these examples. The graphs refer to five Italian regions (Abruzzi, Latium, Marches, Piedmont, and Trentino Alto Adige). The district map adopted in the recent Italian political elections of 1994 and 1996 was taken as the initial partition.

Each of the following trade-off graphs refers to a given target objective (PE for population equality, C for compactness, and AC for administrative conformity), associated with the shaded node. A target objective is necessary because local search works with just one objective function at a time. For different target functions, the algorithm proceeds in a different way; it finds a different result, and the trade-off evaluation can be very different, too.

Consider the graph in Figure 12.13(a). The target objective is population equality; in fact, the weight associated with node PE is negative and the absolute decrease is 7% (notice that, since the initial value was 8%, the relative decrease is 87.5%). In our application the problem is formulated in terms of minimization, so that the indexes measure the lack of each objective. Therefore, a high negative variation means a big improvement. In this graph no other index improves, but this is not a rule. In fact, in Figure 12.13(e) compactness slightly increases even if it is not the target objective, while administrative conformity worsens.

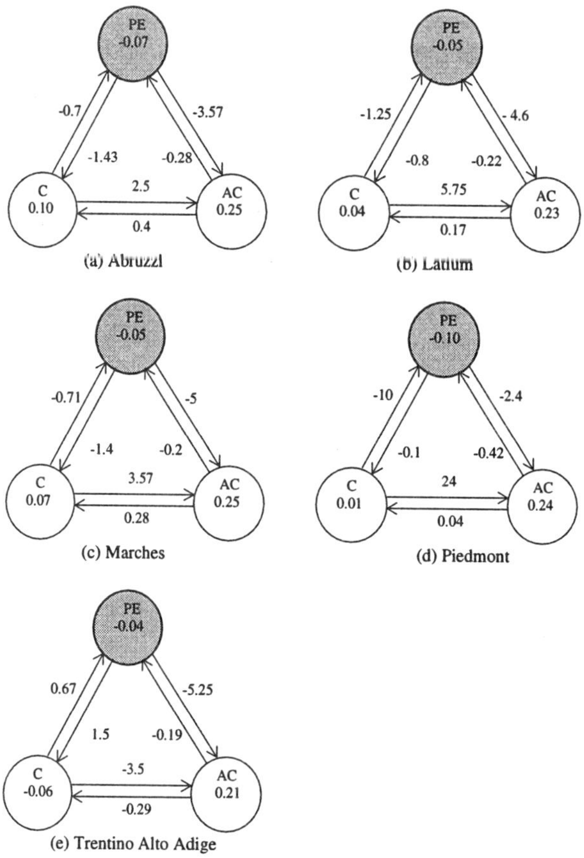

Fig. 12.13. *Trade-off graphs for five Italian regions according to the results of a tabu search application. Target objective: population equality.*

The arcs (C,PE) and (PE,C) have positive weights, which means that the two indexes have varied in the same direction (both have improved). The value 1.5 indicates that a variation of index C corresponds to a variation of one and a half times of PE. Now consider a pair of criteria. If the weights associated with the nodes have the same sign, we are interested in the absolute value of single variations: usually one of them refers to the target value, and the variation is substantial, while the other shows only a small variation. Cases in which two criteria are conflicting are more interesting. Consider, for example, the arc (AC,PE): -0.19 indicates that in order to improve administrative conformity, one must necessarily accept population equality to worsen about $1/5$ of its actual value. In some cases the compromise is not acceptable, but, if conformity is strongly required, it might happen that the final choice is to give up population equality to gain administrative conformity, even if the trade-off is very expensive. Considering the reverse arc (PE,AC), we can also see that a further improvement in population equality corresponds to nearly five times more loss in administrative conformity.

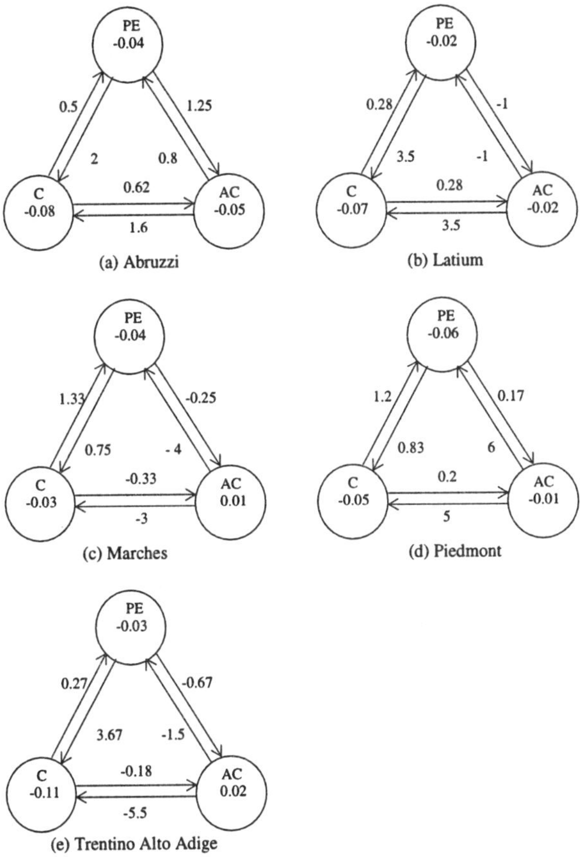

Fig. 12.14. *Trade-off graphs for five Italian regions according to the results of tabu search with mixed objective function.*

Similar trade-off graphs can be built to represent the compromises underlying the solutions obtained when another target objective is considered (compactness or conformity to administrative boundaries) and the remaining two are left unconstrained.

Besides single-objective cases, trade-off graphs can be handy when a mixed objective function, given by a weighted average of the three indicators, is considered. In the specific application we refer to, we have obtained the results shown in Figure 12.14 for the set of weights 0.5, 0.3, and 0.2 for population equality, compactness, and conformity to administrative boundaries, respectively. An interesting result is that a simultaneous decrease of *all* three indicators has been obtained with tabu search for three of the five regions analyzed. This also occurs if other local search algorithms are adopted. Since the initial solution on which the runs where performed was the district map adopted in 1994 and 1996 for the Italian political elections, these results prove that the Italian district map is not Pareto-optimal. Several maps providing better values for the three criteria taken into account can be easily built with the help of an automatic procedure.

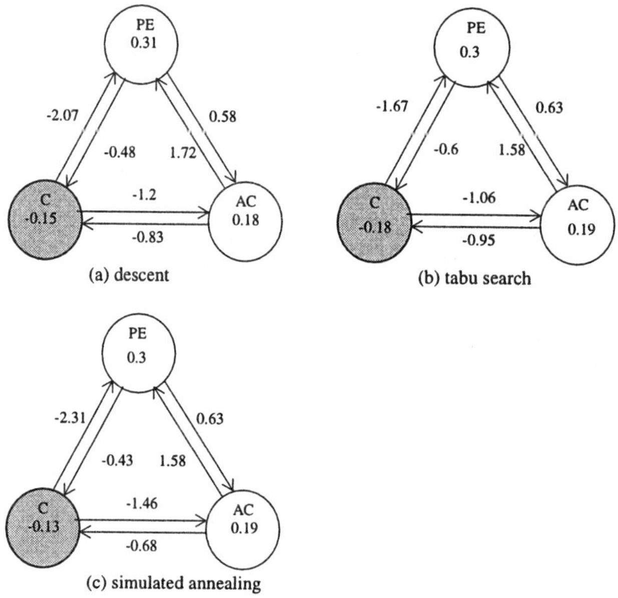

Fig. 12.15. *Trade-off graphs in three different algorithmic approaches. The weights of the nodes are mean values over the five regions. Target objective: compactness.*

Finally, the trade-off graphs in Figure 12.15 are used to compare the quality of the solution obtained with different local search techniques. Here we have reported the results when compactness is the target objective, but since we want to summarize what happens for each algorithm, we consider the average values of the indexes over the five regions under consideration.

In particular, the robustness of trade-offs between the criteria when the algorithm changes can be observed. We can see that in this case gains (or losses) are approximately the same for each algorithm. This means that the trade-offs are very robust; that is, their direction and their intensity is fairly independent from the algorithm used. However, trade-offs are strongly dependent on both the set of weights and the initial solution selected. In a detailed analysis of the set of "good" alternative maps it is therefore highly recommended to consider several initial solutions and a wide range of weights.

Part IV

The Planning and Politics of Electoral Reform: A Retrospective Critical View of a Political Scientist

There are no disciplines, nor branches of knowledge—or rather of research; there are only problems and the need to solve them.

Karl R. Popper, "Realism and the Aim of Science," from the *Postscript to the Logic of Scientific Discovery* (1983)

The first three parts of this book present a quantitative approach for the analysis, evaluation, and design of electoral systems. Among the main objectives of the book, the following must be pointed out:

- the elaboration of a general formal model and the formulation of a code system to describe all existing electoral systems;
- the definition and the comparison between criteria and indicators to evaluate the performance of electoral systems;
- the decomposition of the electoral process into phases, on the basis of which the mentioned criteria may be classified;
- the identification of the virtual or hidden "optimization" process underlying single electoral formulas, considered as a basic tool to enhance our knowledge on the consequences of the corresponding electoral systems;
- the analysis of "mixed" or hybrid systems by simulation to study the variation of a set of performance indicators as a function of the percentage of majority seats introduced in the system;
- the selection and description of a set of criteria, indicators, and methods for political districting, with some experimental reference to the Italian case of 1994.

These objectives are twofold. In the first place, they represent an attempt to use mathematical and statistical tools for a detailed analysis of electoral systems and they respond to the need of precise measurement lacking in this sector that could bring about more knowledge on these topics. The main effort consists of "measuring the *quality* of an electoral system." This somehow leads to the second—and more ambitious—purpose which contains a strong planning perspective. The ambition is that of improving or even optimizing the electoral process or some of its parts (for example, the planning of political districts) and to provide an answer to the question, "How can we select, among all possible electoral systems, the best, the most democratic, the fairest method?"

In this part we will defend the qualitative approach on the basis of the belief, common to all the authors of this book, that the issues can be integrated but not surrogated by mathematical formulation. Our main hesitation is not due to the first purpose but rather to the second, that is, the design of better electoral systems on the basis of objective criteria. In fact, we will try to prove that electoral systems are very far from being conceived as merely technical, objective, and neutral tools since they are fully part of politics. Electoral systems do not only constitute the "rules of the game" in democracy and can therefore affect the political confrontation and the forms of political conflict, but they are in practice the result of the political conflict. In other words, the electoral system cannot always be considered as an independent variable with respect to both the party and political systems. On the contrary, as the following pages will show, in several democracies the recent electoral reform was mainly due to the changes occurring in the party system.

The different language and logic adopted by the quantitative and qualitative approach are the reflection of the different methodology applied. On one side, mathematical modeling and simulation are adopted according to the idea that an adequate formalization of electoral systems can lead to major information on their consequences (in terms of indicators such as government stability, fractionalization, proportionality, etc.) in certain situations and that such information is necessary so that public decision makers can develop better systems. On the other side, history is referred to according to the idea that the most profitable way to increase the knowledge on electoral laws is to analyze their role and mechanism case by case, in time and space, and to compare the different cases to define partial theories capturing the recurrence of the phenomenon and enabling one to formulate forecasts.

Nevertheless, the comparison of these two different approaches is essential to enrich the corresponding background, and the results of our work will depend on how much we have managed to improve the comprehension of electoral systems and their importance in the political process.

Chapter 13

A Difficult Crossroad

Two main theoretical premises are at the basis of the philosophy of research that adopts quantitative methods. The first concerns the theory of the rationality of the actors, and the second relies on the functions of applied social sciences. Therefore, on one side the behavior and the decision strategies of the actors in the political arena are generalized, while, on the other, it is assumed that social sciences can strongly affect life in practice and political sciences in particular can affect political action. A critical discussion of these two premises is necessary to fully understand the objections we discuss.

The theory of the rational actor transfers to the political context economical claims based on utilitarian behavior. In fact, it is assumed "that individuals which act in political arenas are guided by a merely instrumental rationality aimed at the optimal fulfillment of goals which significantly concern their living conditions and that actual social outcomes are due to the interaction between individual components of society and, furthermore, they are governed by collective decision procedures which affect individual preferences" [Martelli, 1989]. One of the most typical conclusions of rational choice theories is "methodological individualism" [Antiseri, 1996], that is, the belief that all social phenomena arise from individual behavior. This means that political actors (voters, leaders, parties, etc.) are always trying to maximize their own profits or benefits (such as nominations, votes, power, and more) or to minimize their own economic and social costs. From a scientific point of view the consequences and the appeal of such an approach are straightforward. By ascribing a rational behavior to the individuals, aiming at the achievement of a maximum utility, and by believing that they are always capable of choosing according to such criterion, the possibility to build up general theories in political science is enormously increased: in fact, once the existing alternatives are known, the choice of the individuals can be foreseen given a fixed situation. The impression that the potential of political research is enhanced is further justified by the fact that sophisticated statistical and measurement procedures can be applied.

Many sectors of political science have developed on the basis of such premises into refined logical-deductive conjectures, among which Downs's analysis of party competition [Downs, 1957], Riker's analysis of political coalitions [Riker, 1962], Arrow's impossibility theorem [Arrow, 1963], Olson's collective action dilemma [Olson, 1965] and Buchanan and Tullock's calculus of consensus [Buchanan and Tullock, 1962]. There is no doubt about the usefulness of the intuitions and the positive contribution of such theories to the development of political sciences, but limits, contradictions, and sometimes nonconvincing results have arisen in more than one case, to the point that Robert Dahl (1996) has stressed that the space dedicated to rational choice in the history of political science is much more "modest" than what its enthusiastic supporters are ready to believe. However, it is not the purpose of these pages to develop an exhaustive debate on such a theme[10]; therefore only a few remarks will be stressed.

The attraction of rational decision theories on social scientists is basically due to the illusion that, by the use of mathematical models, the precision, exactness, and universality generally ascribed to natural sciences can be attained, thus disclaiming the status of "poorer relatives" social sciences are given when compared with natural sciences. It is perfectly legitimate to seek general laws, and no scientist should a priori reject such approach, but it is an error to rely on the pretension of universality to justify a scientific approach and to deny more restricted objectives such as "partial" theories (circumscribed in time and space) that take into account the multitude of cultural, historical, and uncontrollable variables that are involved [Almond, 1990]. Universal theories certainly have more appeal, but (in our opinion) social phenomena cannot be reduced to elementary models based on a limited number of controllable variables.

Moreover, while rational models can be useful to develop forecasts in the analysis of several microprocesses (a typical application could be the analysis of decision strategies in committees) [Sartori, 1987], they are useless in the analysis of political and social macroprocesses and thus in the majority of the issues social sciences address. This confirms Popper's statement according to which the main duty of theoretical social sciences "consists of the definition of the social, and not intentional, consequences of intentional human actions" [Popper, 1969]. In other words, social and political phenomena cannot be considered the result of intentional actions since "some events which occur in social life are not necessarily someone's aim" [Moon, 1975]. The reason is that many events do in fact occur against intentional actions, and an enormous amount of separate individuals participate in the process. This same concept has been stressed more than once in the analysis of macrophenomena such as revolutions: as written by Theda Skocpol (1979), quoting Wendell Phillips, "revolutions are not made, they just happen"; or as noticed by Charles Tilly, the complexity of revolutions is due to the fact that they are "the most visible outcome of many independent processes, in the same way that the variation of the population in a city is determined by the sum of the effects of migrations, birth and death rates" [Tilly, 1975].

In practice, it must be understood that social phenomena are basically un-

[10]For an exhaustive debate the reader may refer to Pappalardo (1989) and Green and Shapiro (1995).

determined since the actors involved do not have a precise and binding set of alternatives but rather a very large number of possible choices: therefore, the outcome cannot be foreseen a priori, although it can be explained a posteriori.

The second premise is based on the conviction that scientific results in the field of political research can be applied to real life, thus contributing to change it and to improve it. The appeal of applying scientific theories is strongly connected to their expectation and forecasting capacity: the more one considers that forecasts are possible, the more one is trustful of the application of political science. From our point of view, caution must be used when discussing the potential of applied political science. In fact, there is no antinomy or "constitutive contrast" [Sartori, 1979] between practical and scientific goals, but, at the same time, we must be aware of

(1) the limits of an applied theory of political science;

(2) the substantial difference between intervention at a micro and a macro level;

(3) the need to rely on the capability of determining the most appropriate modes to obtain one's goals, especially in the more general cases;

(4) therefore, the need of constantly referring to values which, in turn, are necessary to define the objectives of political action;

(5) the danger underlying a "planning attitude" based on the pretext that everything may be reformed as long as one wants to;

(6) the unexpected costs that can arise from every social action and that can become unsustainable in the case of widely applied planning measures [Sartori, 1979; Fisichella, 1985; Pasquino, 1989].

The basis of a strong attitude for applied political sciences is the so-called conditional analysis: if we know that a given event is the outcome of a series of circumstances, we also know that to repeat the event in a different context, the same circumstances must be reproduced; analogously, if we want to prevent such an event from occurring, we must prevent the conditions from holding. Hence, the analysis of the conditions under which phenomena take place allows us to determine the potential and the limits of external measures.

Another general feeling that moderates any enthusiasm toward the application of political science is motivated by the fact that scientific theories in this field have scarce forecasting possibilities in any case. This is the opinion of Panebianco (1989) when he states that "only negative forecasts are possible in the field of social sciences" telling us what should not be done but giving us no answer on what should be done. This is certainly not the most widely accepted view among political scientists, who are generally favorable to the application of their theories and, furthermore, encouraged by the expectations others (such as governments, parties, etc.) have in this regard. Nevertheless, it cannot be ignored that such a view is enforced by the practical results obtained up to now. The conclusions carried out by Panebianco both on the quality of applied research and on its results are quite discouraging and rather realistic. Applied research is frequently a tool used to justify decisions that have already been made and nonneutral benefits, rather than a tool to influence public policies.

It follows from this brief discussion that political science must suggest, inform, and help to better understand political phenomena to provide aid in decision making by continuously reminding us to be circumspect, cautious, and attentive. Experience proves that the decision outcomes, that is, the political choices, do not draw from the direct action of scientific knowledge, but they are the—usually unexpected—consequence of spontaneous cooperation, compromise, and bargaining processes between constantly competing and changing powers. Science and politics cannot be considered separate and isolated compartments: the scientist is, against his will, totally part of the social and political environment, and even the most impartial technical recommendation is not immune to the accusation of partisan intentions. This consideration leads to a different perspective of science according to which the forecasting capacity generally attributed to social sciences is reduced, and the analysis must take into account the peculiarity of the phenomena and the fact that political and social actors cannot always be considered rational decision makers.

Chapter 14

The Planning and Politics of Electoral Reform

14.1 Applying Political Theories: Electoral Engineering

Electoral and institutional engineering are privileged test fields for the application of political science. There may be different interventions both in type and in depth, although history proves that major transformations are never decided around a table, but they are the result of extraordinary circumstances (such as the turning from the Fourth to the Fifth Republic in France, which was induced by the Algerian crisis) or the conclusion of gradual and incremental process in the long term. In the following pages, we will restrict our attention to electoral systems. Our purpose is not to discuss in detail the proposals and analysis made in this book but rather to point out which role electoral systems actually play, taking into account both the aim underlying the quantitative approach and the remarks made up to now. The question we would like to answer is, "To which extent can quantitative methods (which are without any doubt useful) actually be used to design electoral systems and contribute to making them neutral tools?"

The evaluation of electoral systems can follow two directions of research. The first one mainly considers the analysis of the political consequences of electoral systems. In such a case, electoral systems are viewed as the subject of change and we focus on *what* they do and *how* they do it. The second is concerned with the analysis of the conditions under which specific electoral systems are implemented, substituted, or changed. In this case, electoral systems are the object of change, and what we are interested in is why and how an electoral system is reformed. These two research lines are clearly connected one to the other, but for the purpose of our analysis they will be treated separately.

14.2 Electoral Systems and Political Parties

It is a well-known fact that when the "effects" of electoral systems are mentioned, three distinct areas of intervention are referred to: the manipulation of the voters' choices, the over-/underrepresentation of parties and the number of parties [Fisichella, 1982]. This means that, putting aside the indirect effects of electoral systems on the political system in general, the real field of their action is restricted to parties and party systems. Such a statement does not exclude the wider effects electoral systems have on the internal compactness of parties, on the types of cooperation and coalitions brought forth, on the relationship between elected candidates and those who vote for them, on the electoral campaigning, on party recruitment, etc.

Duverger's "laws"(1954) (stating that two-party systems are enhanced by plurality and multipartism by double ballot and proportional representation) were the first step toward the definition of a more exhaustive theory and toward a long-lasting debate that has, in the end, adjusted these same claims and proved that parties cannot be considered a simple dependent variable.

In other words, as already shown in the many observations made in this book, the relationship between electoral systems and parties is much more complicated than what Duverger's laws seem to suggest. The core of the problem is represented by three main considerations. The first is that the effects of electoral methods must be considered at two different levels: within districts and at national level. The second refers to the structure of the parties and party systems on which the electoral system acts. Finally, the third concerns the type of effect electoral systems have. The importance in keeping clear these distinct levels of consequences is due to the different—and sometimes contrasting—conclusions made when not taking them into account.

In fact, referring to the first point, effects on single districts do not necessarily coincide with those at national level: for example, plurality tends to restrict to two candidates the competition in a district, but this does not automatically lead to two-party systems. Two-party systems develop only when the same two parties are the main competitors in most of the country's districts. Similarly, the deviation of vote shares from seat shares increases as the size (in terms of seats to assign) of the district increases.

The second point stresses the difference between party systems that are structured and those that are not. We define as "structured" a system that includes organized, widely consolidated, nationwide parties, capable of attracting the main orientations present in society. As mentioned by Sartori (1987), no influence of electoral systems can be detected on party systems in general, unless they are structured. In the best case, this influence will be restricted to the district level. However, when newborn democracies are taken into account, the "first" electoral system can strongly affect the birth and development of the party system.

Eastern Europe provides interesting examples in this direction: in countries where weakly selective systems have been adopted, the participation of a very large number of political groups was encouraged. Take the case of Poland: in the 1991 elections this country used a proportional-based system obtained by combining the Hare–Niemayer method in the districts and assigning remainders at the national level by the Sainte-Laguë method, with an exclusion threshold equal to 5% of the total vote or to at least five seats in the smaller districts. The departing Polish Lower Chamber (*Sejm*) included over 30 parties, among which 11 were assigned only 1 seat [Webb, 1992].

On the contrary, when the electoral system adopted is very selective, the large number of parties and political groups born in the spirit of the newly established pluralism will gradually be discouraged after a couple of total failures in repeated elections. This is the case of Hungary, where the mixed system which is adopted (176 single-member seats, 152 regional seats, and at least 58 national seats) implies that at least 4% of the regional votes must be obtained to participate in the assignment of the regional and national seats [Gabel, 1995].

Even though it is true that in such cases the electoral system worked like a filter, this is not sufficient to claim that it is an independent variable with respect to party systems. In fact, the parties (in spite of their recent formation) can react to the existing situation by operating on the electoral system themselves, by changing it or reforming it. This is proved by the Polish case since the second elections (1993) saw the use of a slightly modified electoral system. Furthermore, the electoral system is only one among a set of many variables affecting the development of the party system and it is not the most important. It can be considered as an additional obstacle (of minor or major size) that the parties must surmount to obtain seats and it is usually not the most difficult. So, which other selective factors exist?

A newborn party, before even dreaming about the electoral system, must tackle many other problems such as the definition of an organizational structure; the definition of a public image to attract supporters; the development of political strategies and proposals; the diffusion of its ideas; the recruitment of funds, candidates, and leaders; and a stable electoral basis. The importance and the weight of these issues is straightforward in Eastern European countries where new parties must compete with the former Communist parties: maintaining and adapting an already existing structure is very different from building a completely new one. If the newborn parties do not address most of these issues, their destiny will be written, whichever the electoral system adopted.

In addition, in this first phase of development, the single parties also contribute to the definition of the new political environment by ordering themselves on a left-to-right political axis. This implies important choices related to the definition of strategies, the political image, the participation in wider political "families" (liberal, confessional, etc.), as well as the relations with the other parties and, therefore, the adherence to coalitions, merges, schisms, the changes in the name of the party, and so on.

Only at the end of a similar process—longer in some experiences (Eastern Europe) than in others (Italy, Germany, and Austria after World War II, but also Spain, Greece, and Portugal in the 1970s)—the party system may be considered stable. The electoral system can certainly condition such a process: in general, if a proportional system is initially adopted and the system is modified at each successive election, the stabilization process will require more time to make sure that no other factors change. In the Polish and Czech cases, this preliminary phase of development coincides with the disaggregation of the large coalitions who were the main actors of the transition (*Solidarność* in Poland and *the Civic Forum* and *Public Against Violence* in Czechoslovak Republic), just as the party systems in Italy and France have a direct link with the dissolution of the antifascist national liberation committees. The fractionalization of the Polish Parliament is due to such circumstance as well as to the proportionality of the electoral system. The proof is given by the fact that the Senate, although elected on a majority system, is characterized by a similar degree of fractionalization. The same holds for the Czechoslovak case in which the Federal Assembly increased from 6 parties in 1990 (the largest representing 170 seats) to 15 parties in 1991 (the largest representing only 43 seats).

The three considerations mentioned above must therefore always be kept in mind to distinguish the actual effect of the electoral system and that of other factors. In general, electoral systems have a relevant effect only on structured party systems, and their effect is rather at the national level than at the district level. Moreover, such effects are noticeable in the long term rather than immediately. In fact, it is only in the long term that the continuous underrepresentation of a party will induce its supporters to vote for others who have a higher probability of winning. On the other hand, this does not necessarily mean the party will disappear: after over 30 years of underrepresentation, the French Communists can still rely on an electorate basis that has also resisted to the collapse of Eastern European communism. Political subcultures (either ideological, ethnic, linguistic, or religious) which are deeply rooted in the society can contradict the underrepresentation effects due to the electoral system.

In any case, the general trends related to the operational functions of the different categories of electoral systems can be identified only in the long term and on structured party systems [Sartori, 1987; Fisichella, 1982]. Experience shows that, *ceteris paribus* and unless opposing conditions exist, majority single-ballot systems encourage bipolar party systems to develop and two-party systems (if they already exist) to stay so, while proportional systems encourage multipartism. The effects of double-ballot (or French) systems are even less clear cut: if the positive effect on the number of parties Duverger claims (which is contradicted, however, by the experience of the Fifth Republic in France) is excluded, some authors claim it has selective effect [Sartori, 1987] and others claim that double ballot does not have any particular effect at all [Fisichella, 1988]. However, a negative effect on antisystem parties has been asserted and is continuously proved by the underrepresentation of the Parti Communiste de France and, more recently, that of Le Pen's Front National. The degree of the penalty inflicted is related to the selectivity of the rule (exclusion threshold) defining the access to the second ballot.

14.3 Designing Electoral Laws: The Four Main Processes of Reform

Up to now we have analyzed the consequences electoral systems have on party systems: electoral systems do have an effect, but it is limited to a number of circumstances and conditions which make Duverger's laws unsuitable today. Theories considering electoral laws as the main factor for the development and the characterization of party systems are, in our opinion, unsustainable.

However, when the modes and phases leading to electoral reform are considered, the point of view may be radically different. Clearly, we only refer to electoral "revolutions" and not to those small and minor modifications that are quite frequent in the history of Western democracies, such as adjusting district boundaries, modifying exclusion thresholds, etc. The establishment of a new electoral system is certainly based on the knowledge and perception of its consequences, but the problem is the extent to which such knowledge is, in practice, effective. Therefore, in the following pages, we will attempt to explain if and in which measure the choice of an electoral system is the result of conscious and rational political choices.

From the point of view we are now considering, parties play the role of independent variables. The entire political development of Western Europe proves that, once party systems are structured and once they have developed stable connections with their electors, it is very difficult for a new electoral system to change, alone, the current establishment. Moreover, the history of Western democracies is essentially the history of political parties and movements which fight—with success (recall the evolution of the working class in Europe)—the wall built against them. We agree with Lipset and Rokkan (1967) when they write that "in most cases it makes little sense to treat electoral systems as independent variables and party systems as dependent," mainly because electoral laws are usually a direct consequence of political strategies: parties principally aim at strengthening themselves with (new) electoral laws.

Such an approach does not only reverse the cause-and-effect relationship between electoral and party systems, but it also stresses the need of identifying other, more significant, variables. For example, while attempting to reconstruct and to explain the origins of parties and party systems in Western Europe, Rokkan [Lipset and Rokkan, 1967; Rokkan, 1970] calls our attention to the role of cleavages: the political contrast in Europe at the end of the nineteenth and beginning of the twentieth century developed on the basis of conflicts that were the result of a specific historical process. This process was common to all countries, although it presented different features according to the different national context. The main contrapositions involved the conflict between center and periphery, between government and the church, between city and countryside, between owners and workers, all contributing to a fight lasting several dozen years. In this context, electoral systems appear to be more an additional variable than a determining one.

Some political scientists prefer to point out the importance of institutional variables. This is the opinion of Lijphart (1994), who claims that "presidentialism tends to discourage multipartism." This also refers to semipresidential systems, like those in use in France, Finland, and Portugal, as well as semiparliamentary systems, such as those in Austria, Iceland, and Ireland, for which "it may be hypothesized to have similar effects." In such a case, supposing that the electoral system had a negative effect on the number of parties in the Fifth French Republic, we must try to understand to what extent such effect is due to the electoral system rather than to institutional variables, such as the semipresidential government.

A tangled knot which must still be unraveled concerns how, why, and who decides the electoral system in a given context: how and why is a particular system chosen and who decides and on which basis and with which aim? A comprehensive answer to these questions must take into account the many different cases constituting the history of democracy. We think that four main processes, which in real life are continuously mingled and combined, can include most of them.

The first process characterizes newborn democracies. In this case the electoral law is usually the outcome of tradition and national heritage. The development of the majority system (with or without single-member districts) is based on the idea of territorial representation [Bogdanor and Butler, 1983]. Therefore, it tends to be adopted in countries that prove a strong tradition in representation even before political parties were actually developed. Rokkan (1970) points out that the plurality system was effective in England even in the Middle Ages when two knights were elected to represent each county and two civilians to represent the boroughs. More or less the same system was adopted in the United States (in a single-member version) shortly after. Other majority and multiballot methods were widely spread among territories dominated by the Catholic church and were still in force at the end of the nineteenth century in many countries (from France, to Italy, to the German Reich, to Austria and the Netherlands).

The origins of proportional representation are very different. In many cases it followed from the need to protect ethnic, cultural, or religious minorities in the framework of "a territorial consolidation strategy." In fact, leaving such minorities out from representation would have endangered the state building process. This is why proportional representation was born in countries such as Denmark (1855), the Swiss Cantons (1891), Belgium (1899), Moravia (1905), and Finland (1906). In other cases, however, the development of proportional representation is strictly linked to the conflict between two specific political powers: the newborn working class wanted to lower the representation threshold in order to access the legislative assemblies and the older traditional parties, threatened by such perspective, suggested the proportional system in order to protect their position with respect to the new vague of electors established by universal suffrage [Rokkan, 1970]. Such an explanation holds for Italy in 1919 as well, where the pressure of the new Catholic electorate required representation. Nevertheless, proportional representation develops to represent opinions, political parties, or aggregations, rather than territories, according to the fact the cost of nonrepresenting or underrepresenting certain powers would be considerably higher than the benefits.

The situation is totally different when an existing electoral law is radically modified. In such a situation, the reform can follow three different processes: (1) it is carried out by imposition; (2) it is the result of a bargaining process; (3) it is decided by a referendum. In the first case, a group of predominant actors manages to impose the use of a new electoral law in a relatively short time with the help of a solid majority in Parliament or of a particular phase of transition. The electoral reform is not the outcome of an agreement between the majority and the opposition but the result of a (nearly) one-way decision. Examples of such kind of reform are the so-called Acerbo law in Italy in 1923 and the double reform which took place in France in the 1980s. In 1986 in France, while its popularity was decreasing, the Socialist party changed the electoral law. In spite of the fact that the new law was intended to provide an advantage to this party, the center-right parties won and immediately put back into place the old electoral law (double ballot).

A similar one-way decision process occurred in the passage from the Fourth to the Fifth Republic in France again. A necessary (but very uncommon) condition for such a process is a political atmosphere in which the electoral law is not considered particularly different from any other law. On the contrary, when there is the widespread conviction that the electoral law can modify the current power relationships, electoral reform is not undertaken without pain. A similar attempt of electoral reform was carried out in Italy in 1953 (the so-called *legge truffa* or fraud law), but it failed thanks to the opposition's campaign.

Other electoral reforms follow a more difficult and longer path based on the bargaining process between opposing political powers which eventually find an agreement. This is, in part, the process undertaken for the reform of the Japanese system in 1994 which brought the single nontransferable vote to an end in favor of a mixed majority/proportional system. This is also the case of many postcommunist regimes in Eastern Europe (and, with small variations, that of postfascist Italy). Of the many phases leading to the establishment of a democracy (the transition, the rise of political parties, the bargaining process between the new and old actors, first free elections, new constitution, etc.), the bargaining process is the most crucial and generally decides on the electoral system to be adopted for the first democratic election. The outcome of such a process obviously depends on the relative power of the actors: in Hungary the communist government supported the reform and was stronger than the newborn opposition; in the Czechoslovak Republic the communists were totally against the reform but, in the end, weaker than the combative opposition.

The third reform process is that in which a referendum constitutes the main factor of change by forcing the political parties to admit the existence of a widely spread public opinion supporting the substitution of the old electoral system. This is what happened in Italy in 1993 when the Parliament was induced to pass a new electoral law given the outcome of a referendum which abrogated a few articles of the current law and was strongly advocated by a large political movement. On the other hand, in New Zealand, the referendum had already been established as the actual tool to decide on the electoral law, and it was supported even by those parties who were against reform and who thought that the citizens would have voted for the status quo. These two referendums have decided the characteristics of the new electoral systems, substituting the previous proportional system with the current mixed 75% plurality system in Italy, and the plurality system with a mixed-proportional system in New Zealand.

14.4 What Are the Reasons for Electoral Reform? Three Recent Examples

In well-consolidated democracies the radical reform of electoral laws is a rare event. However, at the beginning of the 1990s, major reforms are reported in three different democratic countries: New Zealand, Italy, and Japan. The debate on electoral systems is further animated by Eastern European countries where reforms or minor modifications of the newborn electoral system are necessary (for example, in Poland). Besides the motivations put forth by the political actors in favor of reform (such as encouraging a more functional political system, stability, governability, or enlarging the representation basis), what are the real causes of a reform? What leads a political class to discuss the basic "rules of the game" in force for several decades?

The process undertaken by traditional democracies such as those mentioned above suggests various considerations. Clearly, in all three of these countries a general feeling that the current system was inadequate grew in time. Not only did several political parties demand a change, but many "experts," party leaders, and other institutional representatives call for the growing need to solve such a problem. In Italy and in Japan, recent news of political corruption strongly perturbing the national scene was associated with the old electoral law, which was further criticized in Italy for encouraging party fractionalization and unstable governments. In New Zealand, the reform was advanced not only by the new political actors, whose role was increasing in spite of the fact they were penalized by the plurality seat allocation method, but by the traditional parties as well. The old electoral system had in fact allowed the National Party to win the majority of the seats in Parliament twice in a row (in 1978 and in 1981), although the Labour Party had received a larger number of votes [Mackie and Rose, 1991]. Traditional parties were also worried about the growing distrust in politics in general.

However, both in Italy and in New Zealand, the debate on the electoral reform could have remained unsolved if it had not resorted to a decisive referendum. In fact, although specific committees were appointed (the Bicameral Commission instituted in 1983 in Italy and the Royal Commission established in New Zealand in 1984), at the end of the 1980s the progress toward reform was still unsatisfactory. In Italy, in particular, although everyone agreed on the necessity of a reform, no proposal succeeded in gathering the consent of the majority. Not only did it seem difficult to find an agreement, but the political parties were jumping from one proposal to another according to the place, person, and convenience of the moment [Pappalardo, 1994]. All the actors involved were demanding a reform (it would have been unpopular not to), but their behavior was often procrastinating and their proposals contradictory if not totally fanciful. The parties also proved to be scarcely convinced and scarcely informed.

The point is that an electoral reform produced by no external factor but just as the result of an agreement between political parties is very difficult to achieve for at least two reasons. First, the nature of electoral laws is itself a barrier to reform. Electoral systems determine the key rules of competition in democracies and it is difficult to change these rules in the middle of the "game": some "players" will always fight against the change. Electoral laws cannot be considered just like other laws and they are perceived as a strategic issue in politics since they can affect, together with other factors, representation [Taagepera and Shugart, 1989]. Second, finding an agreement in this context is usually difficult because of the large number of different actors involved (parties, social powers, intraparty components, institutional bodies, mass media, public opinion) and because, in some cases, the actors might use their position on this issue in exchange for something else, they might follow and/or contribute to the infatuations of the public opinion, they might base their proposals on tactics that are totally inconsistent (for example, supporting some position just in order to divide the opponents). Finally, every electoral system has a proper aim, and trying to find a compromise might produce strange and complicated methods that are not understood by the electors and will contribute to widen the gap between the actual purposes and the consequences of a reform.

So, what are the conditions that do force the reform to pass? Generally, these conditions are the pressure due to unpostponable deadlines and the development of a political situation where the continuity of the democratic regime is somehow disturbed. In Italy and in New Zealand the final impulse toward the reform was in fact due to the adoption of a referendum. In Italy, immediately after the referendum had abrogated some of the articles of the existing law for the Senate, a period of unusual efficiency in Parliament led to an intense production of the laws on the elections of communal and province administrative bodies, on the election of the Senate and the Lower Chamber, as well as the rules of electoral campaigning, in just six months, from March to August 1993 [Lanchester, 1994]. In New Zealand two referendums were needed to substitute the plurality system with a mixed-proportional system. The purpose of the first referendum in 1992 was merely to investigate the general attitude toward a reform and the citizens' main preferences. Although the number of electors was much lower than in legislative elections (only 55% against an average 90% in the last 25 years), the outcome was totally in favor of the reform (84.7%) with a strong preference for a German-like system (70.5%). The second referendum in 1993, which took place during the general elections [Levine and Roberts, 1994], confirms the previous results but with a much smaller gap between the two opinions: those in favor of the current plurality system represented about 46.1% while those in favor of a mixed-proportional system were about 53.9%. Moreover, the discontinuity of the electoral rules supported by the results of the referendum was followed by an electoral result which favored continuity, since the conservative government in charge was reelected [Levine and Roberts, 1994].

Japan is a different case: the impulse toward reform was not the result of a referendum but of an extraordinary political event, that is, the constitution of a government led by Hosokawa and supported by a coalition of the parties at the opposition with the exception of the communists, but including the socialists, which are the traditional antagonists of the Liberal Democrats (LDP). The declared purposes of the coalition were to decree the end of the single-party government of the LDP and to pass an important political reform with the symbolic value of proving that Japan really was a changing country. Together with the introduction of a public financing system for the parties, the reform of the electoral law was passed. The law in force allocated 300 Parliament seats in single-member districts with a single transferable vote and the remaining 200 seats proportionally distributed among 11 regions [Shiratori, 1995].

The referendum is used to give an impulse to the stationary situation in Italy and New Zealand on one side, but on the other it is not sufficient to explain the real reasons for reform and it does not answer the question, "Why are electoral rules changed after several decades?" In Japan the reform process did not adopt a referendum. The real common denominator of these reforms is the underlying party system transformation process. In other words, all three countries confirm the thesis of Lipset and Rokkan, according to whom the independent variable in the relationship between party systems and electoral systems is the party system. We infer that *the more a party system presents symptoms of instability and transformation with respect to the past, the more a radical reform of the electoral laws is probable*. In fact, in such a situation, the demand for change increases while the defenders of the status quo decrease. The reform will be embraced by the new political challengers arising from the destabilization of the system and it will be fought by the traditional parties already in crisis and not very keen to abandon the rules that can, in their opinion, guarantee their survival.

In Italy, in addition to the crisis of the traditional parties (made evident by the rapid increase of abstention to vote in the 1980s which reached a peak of 17% in 1992), of party fractionalization [Sani, 1994] and electoral volatility (which is two times greater between 1987 and 1992 with respect to the 1972–1987 period), the inquiries on political corruption (*Tangentopoli*), and the rise of a movement with the aim of substituting the proportional system with a plurality system, are key factors in the reform process. The acceleration of the electoral reform followed the acceleration of the party crisis since delegitimated political powers had to cope with the clear wills the citizens had expressed in the referendum.

In New Zealand, the crisis of the two-party system reached its summit in 1993 when about 30% of the total electors voted for nontraditional parties (see Table 1). However, the crisis had its roots since 1972. Electoral volatility increased during the 1980s (16.7% of the electors changed their vote with respect the past election in 1984 and 19.7% in 1987) and it reached maximum values in 1990 and in 1993 when it was over 20%, twice as much as the average value reported between 1966 and 1987. Several opinion surveys proved widely spread distrust in politics and its traditional institutions, that is, the two traditional parties, the Parliament, and the electoral system [Vowles, 1995; Levine and Roberts, 1994]. Therefore, the electoral reform following the double referendum must also be considered as an attempt to recover the citizens' confidence through the adoption of an electoral system favoring the political readjustment in effect for several years and strongly penalized by the previously adopted plurality system.

Table 1. *Percentage of votes cast for nontraditional parties and corresponding number of seats in New Zealand.**

	1954	1966	1975	1978	1981	1984	1990	1993
% votes	11.1	14.5	12.6	18.5	20.9	20.1	13.7	28.6
No. of seats	0	1	0	1	2	2	1	4

*Source: elaborated on the basis of Vowles (1995).

The symptoms of the Japanese party system instability mainly were due to the increase in party fractionalization and the loss of the LDP majority in Parliament in the 1993 elections, for the first time since 1955. Exhausted by about 40 years of uninterrupted government and by episodes of corruption and schisms, the LDP still remained the major party, but it was forced to govern with a coalition. Such a radical change was announced in the 1990 elections, when the electoral volatility had reached over 13.3% against the 7.6% average value for the last 20 years. This trend was confirmed in 1993 when electoral volatility reached 20.9%. The electoral reform is therefore a consequence of all these factors: the increase of electoral volatility due to the episodes of corruption,[11] the different power relationship between parties, and the rise of new political parties (at least three in 1993) [Shiratori, 1995].

14.5 Conclusions: Three Scenarios for Electoral Reform

The purpose of these pages was to stress the main issues addressed in the debate on electoral systems by political scientists today and to provide the help of a qualitative political science perspective to understand the topic that is analyzed in this book with the support of totally different methodologies. Electoral systems are methods used to transform votes into seats, but they develop in a specific historical and political context. Rather than the outcome of a rational

[11] It was stated that the single transferable vote encouraged the competition between candidates belonging to the same party, thus increasing the need for financial resources for the electoral campaign and exacerbating corruption.

procedure, they are the result of the spontaneous coordination between actors, the intense political debate, the conflict between different powers fighting for representation and for survival, and, finally, they are the result of a compromise. They must be considered "the direct emanation of political strategies." This is why, besides the *technical features* of these methods and their effects, it is essential to understand the *political features* involved in electoral reform and how such a process develops.

We have shown four possible and overlapping reform processes from which three main scenarios stand out to characterize different degrees of democratic development. The first scenario concerns countries that are gradually moving toward an inclusive political system which will, in turn, develop into a modern democracy. The first method adopted to transform votes into seats is generally suggested by history and tradition and/or by the necessity of including ethnic and linguistic minorities for the purpose of building the state. Of course, electoral systems undergo some modifications in time (often demanded by the new political actors) and, above all, the adjustment of electoral district boundaries according to the changes in population, migrations, and urbanization. The delay and the opposition to such changes is mostly the consequence of the effort of the traditional political powers to keep the new ones underrepresented.

The second scenario concerns those countries that come back to a democratic regime after a nondemocratic phase. Tradition and history will affect the new electoral system in this case as well: if the memory of the previous democracy is still alive, continuity with history will be sought and solutions will be proposed to avoid the errors that brought the previous democracy to an end. But at least two additional factors must be taken into account in this case, although they were missing in the case above. The first concerns the features of the transition and, namely, whether democracy is the result of external powers or the result of the revival of political actors. If the latter is true, the electoral system will be the compromise between these powers (such as in many Eastern European countries). The second factor consists of the imitation of more experienced democratic countries: at the end of the twentieth century, new democracies have a large variety of examples to choose from to model their own institutions. Therefore, it is not surprising that in Hungary the Spanish transition after franchism was carefully studied and that many countries were inspired by the stability of the Fifth Republic in France and the Federal Republic in Germany.

The third scenario refers to well-consolidated democracies substituting an electoral law after many decades. Similar to the previous scenario, the many examples of systems adopted in other countries are a pressure on internal decision processes.[12] The main difference is due to the relationship between the party system and the reform. The more the party system is stable, the weaker the impulse to reform: the existing political parties are not interested in modifying the existing law and they are probably even afraid of such change. Unless there is a strong and compact majority advocating the advantages of a new system,

[12] An example is that of the party against proportionality in New Zealand using the slogan "to avoid the Italian chaos" in their campaign.

the guarantee of survival characterizing all organizations (including parties) will prevail. On the contrary, as the symptoms of instability of the party system increase, the probability of a radical reform of the electoral system will increase as well.

The analysis carried out proves that, whatever the scenario taken into account, electoral systems cannot be considered as a merely technical and neutral issue, the result of an unbiased choice. In fact, every electoral reform is a precise political act related to historical events and to the relationship between different political powers. We must agree with Rose (1983, p. 40) when he says "electoral laws cannot be evaluated independently from political criteria and values." Therefore, political scientists must try to combine the guidelines and the models suggested by quantitative analysis to the requirements of the political situation.

Part V

A Short Guide to the Literature

In a 1985 article Arend Lijphart writes, "My first general observation is that the study of electoral systems is undoubtedly the most underdeveloped subject in political science" [Lijphart, 1985a]. This is not the impression we get by analyzing the literature on this subject: it is not an easy task to count and classify the large number of texts, articles in scientific reviews, newspapers, and the large amount of seminars, debates, and congresses which are regularly organized on electoral systems.

Rather than a list of references, what we want to present here is a short guide for readers who might be interested in a further insight on some topic or approach we have investigated in the pages above. The milestones and recent developments of the literature will be stated, as well as some further work worthy of mention that we have encountered in the preparation of this book.

General Remarks

Before attempting any classification of the wide production in this field, we must recall that while the interest in this subject has certainly evolved with democratic representation in the last century, the concern for these problems is very ancient, to the point that refined studies on voting systems can be traced back to the beginning of civilization. Nevertheless, the literature today is still not very homogeneous. In many cases the description of single methods adopted in specific countries is privileged with respect to more analytical and comprehensive studies. Two general approaches seem to prevail: the study of formal models for collective choice problems on one side and the study of the political effects and consequences of electoral systems on the other, which can be classified as follows:

- a statistical-empirical approach, mainly based on the recognition of the different procedures and their comparison in terms of political and social effects, such as the many cross-national and cross-temporal studies and the more general theory of electoral and party systems;
- a mathematical approach, devoted to the technical features of electoral or voting procedures and their underlying properties.

The first approach can be essentially attributed to political scientists. Humfreys (1911) and Black (1958) as well as Duverger (1954) and Lakeman and Lambert (1955) are probably the main historical volumes and they have all been revised in several successive editions. A new line of thought in this field was brought forth by Rae (1967) with *The Political Consequences of Electoral Laws*, which has been the beginning of a still unsolved debate in the books of Rokkan (1970) and Lijphart (1994).

Many specific articles focus on a single aspect of a method or a single political effect. The most frequent topics are government, party fragmentation, disproportionality, electoral fluctuation and abstention, electoral reform, and changes in political regime.

The mathematical approach is very fragmentary. It ranges from technical details of voting procedures and seat allocation methods (which are connected to classical works such as Arrow (1963)) to specific problems such as cyclical majorities (already topic of a controversy between Condorcet and Borda in the XVIII century), manipulation of voting procedures, and many others. Proportional allocation methods are treated in an axiomatic or optimization context by several authors, although the most important have been Balinski and Young (1982a) with their book *Fair Representation* and their many articles spread in the literature of the last twenty years.

The districting problem has stimulated the interest of mathematicians since the 1960s. Authors such as Hess, Weaver, Siegfelatt, Whelan, and Zitlau (1965) and Garfinkel and Nemhauser (1970) have founded the basis for the development of algorithmic procedures to draw district plans with particular attention to the prevention of gerrymandering. Optimality criteria and suitable indexes to measure them are the concern of many studies such as Kaiser (1966); Reock (1961); Young (1988); Taylor (1973); Taagepera (1986); and Niemi, Grofman, Carlucci, and Hofeller (1990).

The authors' interest in the recent Italian electoral reform influenced the contents of this book. Therefore, several Italian references have been consulted and are quoted in this book. They are grouped in a separate section.

Milestones

K. Arrow (1963), *Social Choice and Individual Values* (2nd edition), John Wiley and Sons, New York.

M. L. Balinski and H. P. Young (1982a), *Fair Representation: Meeting the Ideal of One Man One Vote*, Yale University Press, New Haven, CT.

D. Black (1958), *The Theory of Committees and Elections*, Cambridge University Press, Cambridge.

S. J. Brams (1975), *Game Theory and Politics*, The Free Press, Collier Macmillian Publishing, London.

A. Downs (1957), *An Economic Theory of Democracy*, Harper and Row, New York.

M. Duverger (1954), *Les Partis Politiques*, Colin, Paris.

F. Hermens (1951), *Europe between Democracy and Anarchy*, University of Notre Dame, Notre Dame, IN.

V. O. Key (1954), *A Primer in Statistics for Political Scientists*, Cromwell, New York.

E. Lakeman and J. D. Lambert (1955), *Voting in Democracies. A Study of Majority and Proportional Electoral Systems*, Faber and Faber, London.

A. Lijphart (1994), *Electoral Systems and Party Systems. A Study of Twenty-Seven Democracies, 1945–1990*, Oxford University Press, Oxford.

P. Ordeshook (1992), *A Political Theory Primer*, Routledge, New York.

H. Nurmi (1987), *Comparing Voting Systems*, D. Reidel Publishing Company,

Dordrecht, Holland.
D. Rae (1967), *The Political Consequences of Electoral Laws*, Yale University Press, New Haven, CT.
W. Riker (1962), *The Theory of Political Coalitions*, Yale University Press, New Haven, CT.
S. Rokkan (1970), *Citizens, Elections, Parties*, Universitetsforlaget, Oslo.
D. G. Saari (1994), *Geometry of Voting*, Springer, New York.

General Description of Electoral Systems

A. Blais, The classification of electoral systems, *European Journal of Political Research*, 16 (1988), pp. 99–111.
G. W. Cox (1997), *Making Votes Count: Strategic Coordination in the World's Electoral Systems*, Cambridge University Press, Cambridge.
A. Lijphart, The political consequences of electoral laws, 1945–1985, *American Political Science Review*, 84 (1990), pp. 481–496.

Majority and Plurality Methods

S. J. Brams and P. C. Fishburn, Approval voting, *American Political Science Review*, 72 (1978), pp. 831–847.
F. DeMeyer and C. R. Plott, The probability of a cyclical majority, *Econometrica*, 38 (1970), pp. 345–355.
P. C. Fishburn, Dimensions of election procedures: Analyses and comparisions, *Theory and Decision*, 15 (1983).
A. Lijphart, The field of electoral system research: A critical survey, *Electoral Studies*, (1985a), pp. 3–14.
S. Merrill, A comparision of efficiency of multicandidate electoral systems, *American Journal of Political Science*, 28 (1984), pp. 23–47.
J. T. Richelson, A comparative analysis of social choice functions, *Behavioral Science*, 23 (1978), pp. 38–44.

Proportional Methods

M. L. Balinski and H. P. Young, The quota method of apportionment, *American Mathematical Monthly*, 82 (1982b), pp. 701–729.
G. Birkhoff, House monotone apportionment schemes, *Proceedings of the National Academy of Science, USA*, 73 (1976), pp. 684–686.
C. Carter, Some properties of divisor methods for legislative apportionment and proportional representation, *American Political Science Review*, 76 (1982), pp. 575–584.
L. R. Ernst, Apportionment methods for the House of Representatives and the Court challenges, *Management Science*, 4 (1994), pp. 1207–1227.
E. V. Huntington, A new method of apportionment of representatives, *Quarterly Publication of the American Statistical Association*, 17 (1921), pp. 859–870.
T. Ibaraki and N. Katoh (1988), *Resource Allocation Problems: Algorithmic*

Approaches, MIT Press Series, Cambridge.
A. Pennisi (1997), Disproportionality indexes and robustness of proportional allocation methods, *Electoral Studies*, 17 (1998), pp. 3–19.
J. Still, A class of new methods for Congressional apportionment, *SIAM Journal on Applied Mathematics*, 37 (1979), pp. 401–418.

Mixed or Hybrid Systems

A. Blais and L. Massicotte, Mixed electoral systems: An overview, *Representation*, 33 (1996), pp. 15–118.
J. Elklit, The best of both worlds? Danish electoral system in 1915–1920, a comparative perspective, *Electoral Studies*, 11 (1992), pp. 189–205.
A. Lijphart (1986), Trying to have the best of both worlds: Semiproportional and mixed systems, in *Choosing an Electoral System*, R. Taagepera (ed.), Praeger Scientific, New York.

The Effects and Consequences of Electoral Systems

A. Lijphart, The political consequences of electoral laws: 1945–1985, *American Political Science Review*, 84 (1990), pp. 481–496.
S. S. Nilson, The consequences of electoral laws, *European Journal of Political Research*, 2 (1974), pp. 238–290.

Government Stability

V. M. Herman and M. Taylor, Party systems and government stability, *American Political Science Review*, 65 (1971), pp. 28–37.

Proportionality and Disproportionality

G. W. Cox and M. S. Shugart, Comment on Gallagher's proportionality, disproportionality and electoral systems, *Electoral Studies*, 10 (1991), pp. 348–352.
V. Fry and I. McLean, A note on Rose's proportionality index, *Electoral Studies*, 10 (1991), pp. 52–59.
M. Gallagher, Proportionality, disproportionality and electoral systems, *Electoral Studies*, 10 (1991), pp. 33–51.
D. Rae, J. Loosemore and V. J. Hanby, Thresholds of representation and thresholds of exclusion: an analytical note on electoral systems, *Comparative Political Studies*, 3 (1973), pp. 479–488.
J. Loosemore and V. J. Hanby, The theoretical limits of maximum distortion: Some analytic expressions for electoral systems, *British Journal of Political Science*, 11 (1984).

Electoral Participation and Volatility

S. Bartolini and P. Mair (1992), *Identity, Competition and Electoral Availability*, Cambridge University Press, Cambridge.

R. W. Jackman, Political institutions and voter turnout in industrial democracies, *American Political Science Review*, 81 (1987), pp. 405–420.

Number of Parties and Coalitions

J. F. Banzhaf III, Weighted voting doesn't work: A mathematical analysis, *Rutgers Law Review*, 19 (1965), pp. 317–343.
J. Deegan and E. W. Packel, An axiomated family of power indices for simple n-person games, *Public Choice*, 35 (1980), pp. 229–239.
M. Holler, Forming coalitions and measuring voting power, *Political Studies*, 30 (1982), pp. 62–271.
G. Gambarelli, Power indices for political and financial decision making: A review, *Annals of Operations Research*, 51 (1994), pp. 165–173.
M. Laakso and R. Taagepera, Effective number of parties, *Comparative Political Studies*, 12 (1979), pp. 3–27.
N. Leiserson, Coalition in politics: A theorical and empirical study, unpublished Ph.D. thesis (1966), Yale University, New Haven, CT.
N. Leiserson, Factions and coalitions in one-party Japan, *American Political Science Review*, 62 (1968).
J. Molinar, Counting the number of parties: An alternative index, *American Political Science Review*, 85 (1991), pp. 1383–1391.
L. S. Shapley and M. Shubik, A method for evaluating the distribution of power in a committee system, *American Political Science Review*, 48 (1954), pp. 787–792.
P. Warwick, The durability of coalition governments in parliamentary democracies, *Comparative Political Studies*, 11 (1979), pp. 465–497.
J. K. Wildgen, The measurement of hyperfractionalization, *Comparative Political Studies*, 4 (1971), pp. 233–245.

Electoral Reform

G. Almond (1990), *A Discipline Divided. Schools and Sects in Political Science*, Sage, London.
V. Bogdanor and D. Butler, eds. (1983), *Democracy and Elections. Electoral Systems and their Political Consequences*, Cambridge University Press, Cambridge.
J. M. Buchanan and G. Tullock (1962), *The Calculus of Consent. Logical Foundations of Constitutional Democracy*, University of Michigan Press, Ann Arbor.
R. A. Dahl, Reflections on a half century of political science: Lecture given by the winner of the Johan Skytte Prize in political science, *Scandinavian Political Studies*, 19 (1996), pp. 85–94.
M. J. Gabel, The political consequences of electoral laws in the 1990 Hungarian elections, *Comparative Politics*, 27 (1995), pp. 205–214.
S. Levine and N. S. Roberts, The New Zealand electoral referendum and general election of 1993, *Electoral Studies*, 13 (1994), pp. 240–253.
S. M. Lipset and S. Rokkan (1967), Cleavages, structures, party systems, and

voter alignments: An introduction, in *Party Systems and Voter Alignments: Cross National Perspectives,* S. M. Lipset and S. Rokkan (eds.), The Free Press, New York, pp. 1–64.

T. T. Mackie and R. Rose (1991), *The International Almanac of Electoral History,* Macmillan, London.

D. J. Moon (1975), The logic of political inquiry: A synthesis of opposed perspectives, in *Handbook of Political Science, Vol. 1—Political Science: Scope and Theory, and Vol. 3—Macropolitical Theory,* Greenstein and Polsby (eds.), Addison-Wesley, Reading, MA, pp. 131–228.

M. Olson (1965), *The Logic of Collective Action,* Harvard University Press, Cambridge, MA.

K. R. Popper (1969), *Conjectures and Refutations,* Routledge and Kegan Paul, London.

R. Rose (1983), Elections and electoral systems: Choices and alternatives, in *Democracy and Elections. Electoral Systems and Their Political Consequences,* Bogdanor and Butler (eds.), Cambridge University Press, Cambridge, pp. 20–45.

R. Shiratori, The politics of electoral reform in Japan, *International Political Science Review,* 16 (1995), pp. 79–94.

T. Skocpol (1979), *States and Social Revolution. A Comparative Analysis of France, Russia and China,* Cambridge University Press, Cambridge.

C. Tilly (1975), Revolutions and collective violence, in *Handbook of Political Science, Vol. 1—Political Science: Scope and Theory, and Vol. 3—Macropolitical Theory,* Greenstein and Polsby (eds.), Addison-Wesley, Reading, MA.

J. Vowles, The politics of electoral reform in New Zealand, *International Political Science Review,* 16 (1995), pp. 95–115.

W. L. Webb, The Polish general election of 1991, *Electoral Studies,* 11 (1992), pp. 166–170.

Political Districting

M. Altman, Is automation the answer?—The computational complexity of automated redistricting, *Rutgers Computer and Technology Law Journal,* 23 (1997), pp. 81–142.

L. D. Bodin, A district experiment with a clustering algorithm, *Annals of the New York Academy of Sciences,* (1973), pp. 209–214.

J. M. Bourjolly, G. Laporte, and J. M. Rousseau, Découpage electoral automatisé: Application a l'Ile de Montréal, *INFOR,* 19 (1981), pp. 113–124.

M. H. Browdy, Simulated annealing: An improved computer model for political redistricting, *Yale Law & Policy Review,* 8 (1990), pp. 163–178.

N. M. Bussamra, P. M. França, and N. G. Sosa, (1996). Legislative districting by heuristic methods, *AIRO 96 Procedings,* pp. 640–641.

V. Cerny (1985), A thermodynamical approach to the travelling salesman problem: An efficient simulation algorithm, *Journal of Optimization Theory and Applications,* 45 (1985), pp. 41–51.

C. W. Chance (1965), *Political Studies: Number 2—Representation and Reap-*

portionment, Dept. of Political Science, Ohio State University, Columbus.
R. F. Deckro, Multiple objective districting: A general heuristic approach using multiple criteria, *Operational Research Quarterly*, 28 (1979), pp. 953–961.
R. J. Dixon and E. Plischke (1950), *American Government: Basic Documents and Materials*, Van Nostrand, New York.
L. Forrest, Apportionment by computer, *American Behavioral Science*, 7 (1964).
R. S. Garfinkel and G. L. Nemhauser, Optimal political districting by implicit enumeration techniques, *Management Science*, 16 (1970), pp. 495–508.
F. Glover, Tabu search—Part I, *ORSA Journal on Computing*, 1 (1989), pp. 190–206.
F. Glover, Tabu search—Part II, *ORSA Journal on Computing*, 2 (1990), pp. 4–32.
B. Grofman and T. Hofeller (1990), Comparing the compactness of California Congressional districts under three different plans: 1980, 1982 and 1984, in *Toward Fair and Effective Representation*, B. Grofman (ed.), Agathon, New York.
C. Harris Jr., A scientific method of districting, *Behavioural Science*, 9 (1964), pp. 219–225.
W. Hess, J. B. Weaver, H. J. Siegfelatt, J. N. Whelan, and P. A. Zitlau, Non partisan political redistricting by computer, *Operations Research*, 13 (1965), pp. 998–1006.
M. Hojati, Optimal political districting, *Computer and Operations Research*, 23 (1996), pp. 1147–1161.
D. L. Horn, C. R. Hampton, and A. J. Vandenberg, Practical application of district compactness, *Political Geography*, 12 (1993), pp. 103–120.
T. C. Hu, A. B. Kahng, and C. W. A. Tsao, Old bachelor acceptance: A new class of non-monotone threshold accepting methods, *ORSA Journal on Computing*, 7 (1995), pp. 417–425.
H. F. Kaiser, A measure of the population quality of legislative apportionment, *American Political Science Review*, 64 (1966), pp. 208–215.
S. Kirkpatrick, C. D. Gelatt Jr., and M. P. Vecchi, Optimization by simulated annealing, *Science*, 220 (1983), pp. 671–680.
J. M. Liitschwager, The IOWA redistricting system, *Annals of New York Academy of Sciences*, 219 (1973), pp. 221–235.
A. Mehrotra, E. L. Johnson, and G. L. Nemhauser, An optimization based heuristic for political districting, *Management Science*, 44 (1990), pp. 1100–1110.
G. Mills, The determination of local government electoral boundaries, *Operational Research Quarterly*, (1967), pp. 243–255.
R. G. Niemi, B. Grofman, C. Carlucci, and T. Hofeller, Measuring compactness and the role of a compactness standard in a test for partisan and racial gerrymandering, *Journal of Politics*, 52 (1990), pp. 1155–1182.
B. Nygreen, European Assembly constituencies for Wales. Comparing of methods for solving a political districting problem, *Mathematical Programming*, 42 (1988), pp. 159–169.
L. Papayanopoulos, Quantitative principles underlying apportionment methods, *Annals of the New York Academy of Sciences*, 219 (1973), pp. 181–191.

E. C. Reock Jr., Measuring compactness as a requirement of legislative apportionment, *Midwest Journal of Political Science*, 5 (1961), pp. 70–74.
I. M. L. Robertson, The delimitation of local Government electoral areas in Scotland: A semi-automated approach, *Journal of Operational Research Society*, 33 (1982), pp. 517-525.
G. Schubert and C. Press, Measuring malapportionment, *American Political Science Review*, 58 (1964), pp. 302–327.
M. Shugart, Two effects of district magnitude: Venezuela as a crucial experiment, *European Journal of Political Research*, 15 (1989), pp. 353–364.
J. E. Schwartzberg, Reapportionment, gerrymanders, and the notion of compactness, *Minnesota Law Review*, 50 (1966), pp. 443–452.
R. Taagepera (1986), The effects of district magnitude and properties of two-seat districts, in *Choosing an Electoral System*, R. Taagepera (ed.), Praeger Scientific, New York, pp. 91–101.
P. G. Taylor, A new shape measure for evaluating electoral district patterns, *American Political Science Review*, 67 (1973), pp. 947–950.
W. Vickery, On the prevention of gerrymandering, *Political Science Quarterly*, 76 (1961), pp. 105–110.
H. P. Young, Measuring compactness of legislative districts, *Legislative Studies Quarterly*, 13 (1988), pp. 105–111.

Italian References

D. Antiseri (1996), *Trattato di Metodologia delle Scienze Sociali*, UTET, Torino.
F. Arcese, M. G. Battista, O. Biasi, M. Lucertini, and B. Simeone, Un modello multicriterio per la distrettizzazione elettorale: la metodologia C.A.P.I.R.E. A.D.E.N., unpublished manuscript, Roma 1992.
C. Bernardi and M. Menghini, Sistemi elettorali proporzionali: la "soluzione" italiana, *Bollettino UMI*, 7 (1990), pp. 271–293.
R. D'Alimonte and A. Chiaramonte, Il nuovo sistema elettorale italiano: quali opportunità, *Rivista Italiana di Scienza della Politica*, 3 (1993).
D. Fisichella (1982), *Elezioni e Democrazia. Un'Analisi Comparata*, Il Mulino, Bologna.
D. Fisichella, eds. (1985), *Metodo Scientifico e Ricerca Politica*, La Nuova Italia Scientifica, Roma.
D. Fisichella (1988), *Lineamenti di Scienza Politica. Concetti, Problemi, Teorie*, La Nuova Italia Scientifica, Roma.
A. Giannuli (1992), *Tutto quello che vorreste sapere sulle leggi elettorali. Vademecum "Referendum,"* I Libri dell'Altritalia, Libera Informazione Editrice.
P. Green and I. Shapiro, Teoria della scelta razionale e scelta politica: un incontro con pochi frutti?, *Rivista Italiana di Scienza Politica*, 25 (1995), pp. 51–89.
P. Grilli di Cortona, Seconda repubblica o Prima repubblica-bis? La transizione italiana in prospettiva comparata, *Quaderni di Scienza Politica*, 3 (1995), pp. 469–494.
F. Lanchester (1981), *Sistemi Elettorali e Forme di Governo*, Il Mulino, Bologna.
F. Lanchester, L'innovazione istituzionale nella crisi di regime, *Associazione per*

gli studi e le ricerche parlamentari, Milano, Giuffrè Quaderno, 4 (1994).
P. Martelli (1989), Teorie della scelta razionale, in *L'Analisi della Politica*, A. Panebianco (ed.), Il Mulino, Bologna, pp. 159–192.
A. Panebianco (1989), Le scienze sociali e i limiti dell'illuminismo applicato, in *L'Analisi della Politica*, A. Panebianco (ed.), Il Mulino, Bologna, pp. 563–596.
A. Pappalardo (1989), L'analisi economica della politica, in *L'Analisi della Politica*, A. Panebianco (ed.), Il Mulino, Bologna, pp. 193–216.
A. Pappalardo, La nuova legge elettorale in Parlamento: chi, come e perché, *Rivista Italiana di Scienza Politica*, 24 (1994), pp. 287–310.
G. Pasquino (1989), La scienza politica applicata: l'ingegneria politica, in *L'Analisi della Politica*, A. Panebianco (ed.), Il Mulino, Bologna, pp. 547–561.
A. Pennisi (1996), Formule elettorali proporzionali e ottimizzazione di funzioni Schur-convesse (extended abstract), *AIRO* 96 *Procedings*, pp. 630–633.
A. Pennisi (1997), Uno studio dei sistemi elettorali misti (working paper).
F. Ricca (1996), Algoritmi di Ricerca Locale per la Distrettizzazione Elettorale (extended abstract), *AIRO* 96 *Procedings*, pp. 634–637.
F. Ricca and B. Simeone, Political Districting: Traps, Criteria, Algorithms, and Trade-offs, *Ricerca Operativa*, 27 (1997), pp. 81–119.
G. Sani (1994), Il verdetto del 1992, in *La Rivoluzione Elettorale. L'Italia tra la Prima e la Seconda Repubblica*, R. Mannheimer and G. Sani (eds.), Anabasi, Milano, pp. 37–70.
G. Sartori (1987), *Elementi di Teoria Politica*, Il Mulino, Bologna.
G. Sartori (1979), *La Politica. Logica e Metodo in Scienze Sociali*, Milano, Sugarco.
G. Sartori (1997), *Ingegneria Constituzionale Comparata*, Il Mulino, Bologna.
R. Scozzafava, Sistemi elettorali: miti e paradossi della proporzionalità, *Studi Parlamentari e di Politica Costituzionale*, 88 (1990), pp. 5–9.
L. Tentoni (1991), *Gli Strumenti per Cambiare*, Acropoli, Roma.
R. Vacca (1994), *La Qualità Globale*, Sperling & Kupfer, Milano.
O. Vitali (1995), *Il Terremoto Politico del* 1994. *Dal Governo Berlusconi alla Dissociazione di Bossi*, Viviani, Roma.

General Mathematical References

D. van Dalen, R. C. Doets, and H. C. M. de Swart (1978), *Sets, Naive, Axiomatic, and Applied*, Pergamon Press, Oxford.
M. R. Garey and D. S. Johnson (1979), *Computers and Intractability. A Guide to the Theory of NP-Completeness*, Freeman, San Francisco.
A. W. Marshall and I. Olkin (1979), *Inequalities: Theory of Majorization and Its Applications*, Academic Press, New York.
G. L. Nemhauser and L. A. Wolsey (1988), *Integer and Combinatorial Optimization*, John Wiley, New York.
R. J. Wilson (1979), *Introduction to Graph Theory* (2nd edition), Longman, London.

Other References

J. J. Bartholdi, C. A. Tovey, and M. A. Trick (1989), How hard is it to control an election?, Report n.89569-OR, Institut für Operations Research, Bonn University.

P. M. Camerini, M. Conforti, and D. Naddef, Some easily solvable nonlinear integer problems, *Ricerca Operativa*, 50 (1989), pp. 11–25.

L. Cooper, Location-allocation problems, *Operations Research*, 11 (1963), pp. 331–343.

H. Dalton, The measurement of inequality of incomes, *Economics Journal*, 30 (1920), pp. 348 361.

E. W. Forgy, Cluster analysis of multivariate data: Efficiency vs. interpretability of classification, *Biometric Soc. Meetings*, Riverside, CA, 1965.

H. W. Gottinger, Choice and complexity, *Mathematical Social Sciences*, 14 (1987) pp. 1–17.

R. L. Graham, E. L. Lawler, J. K. Lenstra, and A. H. G. Rinnooy Kan, Optimization and approximation in deterministic sequencing and scheduling: A survey, *Annals of Discrete Mathematics*, 5 (1979), pp. 287–326.

G. Gross, A class of discrete type minimization problems, RM-1644 (1956), RAND Corporation.

A. A. Kuehn and M. Hamburger, A heuristic program for locating warehouses, *Management Science*, 9 (1963), pp. 643–666.

G. H. Hardy, J. E. Littlewood, and G. Polya, Some simple inequalities satisfied by convex functions, *Messenger Mathematics*, 58 (1929), pp. 145–152.

M. O. Lorenz, Methods of measuring concentration of wealth, *Journal of American Statistical Association*, 9 (1905), pp. 209–219.

J. B. MacQueen, Some methods for classification and analysis of multivariate observations, *Proceedings of the 5th Symposium on Mathematics, Statistics and Probability*, Berkeley, 1 (1967), pp. 281–297, AD 669871, University of California Press, Berkeley.

F. E. Maranzana, On the location of supply points to minimize transport costs, *Operational Research Quarterly*, 15 (1964), pp. 261–270.

H. P. Young, On dividing an amount according to individual claims or liabilities, *Mathematics of Operations Research*, 12 (1987), pp. 398–414.

List of Specialized Journals

The list of journals dedicated to electoral systems is not very long, although a rather long list of journals ranging from political sciences to mathematical applications dealing with the analysis of electoral procedures exists. The following provide a good starting point:

- Electoral Studies
- Legislative Studies Quarterly
- Pouvoirs–Revue Française d'Etudes Constitutionelles et Politiques

- Representation–Journal of Electoral Record and Comment

and several articles on electoral engineering can be found in

- American Journal of Political Science
- American Political Science Review
- Behavioral Science
- British Journal of Political Science
- Comparative Political Studies
- European Journal of Political Science
- International Political Science Review
- Mathematical Social Studies
- Political Studies
- Public Choice
- Rivista Italiana di Scienza Politica
- Theory and Decision

Web Sites

Now there are several web sites containing all sorts of information on electoral systems. In addition to the temporary government sites with the results of recent elections of several countries, we have found databases and archives with historical results. This is a limited list of sites we suggest:
http://www.barnsdle.demon.co.uk/vote
http://dodgson.ucsd.edu/lij/
http://www.ifes.org
http://igc.apc.org/cvd
http://www.mq.edu.au/hpp/Ockham/67xan2.html
http://www-vdc.fas.harvard.edu/staff/micah_altman
http://www.rev.net/~aloe/district/

Index